ICONS

Industrial Design A-Z

Front Cover: *JVC Nivico 3240 GM* television produced by the Victor Company of Japan, 1970. Die Neue Sammlung, State Museum of Applied Arts and Design, Munich. Photo: Angela Bröhan, Munich.
Back Cover: *Silver Streak*, Model 1038 electric iron manufactured by Saunders Machine & Tool Corporation, 1942–46. Designed by the Corning Glass Works. Photo: © Christie's Images Ltd.2000.
Page 4/5: *Oyster*, multipurpose storage containers for Tupperware, 1990s

Hohenzollernring 53, D–50672 Köln
www.taschen.com

Editorial coordination: Julia Krumhauer, Cologne
Production: Martina Ciborowius, Cologne

Printed in Italy
ISBN 3-8228-2426-7

Industrial Design A-Z

Charlotte & Peter Fiell

KÖLN LONDON LOS ANGELES MADRID PARIS TOKYO

CONTENTS

"Utility is one of the principal sources of beauty ... the fitness of any system or machine to produce the end for which it was intended, bestows a certain propriety and beauty upon the whole, and renders the very thought and contemplation of it agreeable."
Adam Smith, 1759

For over 200 years, the products of mechanized industrial production have shaped our material culture, influenced world economies and affected the quality of our environment and daily lives. From consumer goods and packaging to transportation systems and production equipment, industrial products encompass an extraordinary range of functions, techniques, attitudes, ideas and values and are a means through which we experience and perceive the world around us.

The nature of industrial products and how they come to be is determined by an ever more complex process of design that is itself subject to many different influences and factors. Not least of these are the constraints imposed by the social, economic, political, cultural, organizational, and commercial contexts within which new products are developed, and the character, thinking and creative abilities of the individual designers or teams of designers, aligned specialists and manufacturers involved in their realization.

Industrial design – the conception and planning of products for multiple reproduction – is a creative and inventive process concerned with the synthesis of such instrumental factors as engineering, technology, materials and aesthetics into machine-producible solutions that balance all user needs and desires within technical and social constraints. Engineering – the application of scientific principles to the design and construction of structures, machines, apparatus or manufacturing processes – is an essential and defining aspect of industrial design. While both disciplines are concerned with finding optimum solutions to specific problems, the primary distinguishing characteristic of industrial design is its concern for aesthetics.

The origins of industrial design can be traced back to the Industrial Revolution which began in Great Britain in the mid-18th century, and which heralded the era of mechanization. Prior to this, objects were craft-produced, whereby both the conception and the manufacture of an object were the work of a single individual. With the development of new industrial manufacturing processes and the division of labour, design (conception and planning) was progressively separated from the act of making. At this early stage, however, design had no intellectual, theoretical or philosophical foundation and was considered just one of the many interrelated aspects of mechanical production. Thus the industrial goods of the years up to the 19th century were created by specialists from the technical, materials and production spheres rather than by an industrial designer. Towards the end of the 19th century, however, manufacturers began to realise that they could gain a critical competitive advantage by improving the constructional integrity and aesthetic appearance of their products. As a consequence, they began to in-

vite specialists from other spheres– most notably, architects – to contribute to the design process.
Industrial design subsequently became a fully-fledged discipline in the early 20th century, when design theory was integrated into industrial methods of production. Among the first professional industrial design practitioners was the German architect Peter Behrens, who was recruited by AEG in 1907 to improve the company's products and corporate identity. Since then, industrial design has become an increasingly important factor in the success of industrial products and the companies that manufacture them.
This survey of industrial design encompasses all aspects of the subject – from heavy industrial products to office equipment, and kitchen appliances to aerospace vehicles – and is intended as a companion work to our earlier *Design of the 20th Century*. The manufacturers, industrial design consultancies, individual designers and inventors we feature – in entries organized in alphabetical order – are those who have developed some of the most important and influential products and technologies of the last 200 years. Their innovative solutions strike the best possible balance between the intellectual, functional, emotional, aesthetic and ethical expectations of the user/consumer and the influences and factors bearing upon the design process. Cross-references appear in the text in bold type so as to reveal the many illuminating interrelationships between designers, design consultancies and manufacturers.
As more and more countries are drawn into the global free-market economy, so industrial design has become an increasingly vital means of competing on a global scale. By reflecting on the history of industrial design and focusing in particular on its successes, this book hopes to demonstrate how industrial design has consistently sought to de-mystify technology and to deliver it in accessible forms to the greatest number of people.
A further aim of the book is to highlight the phenomenal extent to which manufacturing companies themselves have shaped the history of industrial design. Without their willingness to risk the necessary and sometimes massive investments demanded by the development of new products, there would be very little industrial design. Innovative and commercially-motivated yet socially-minded manufacturing companies are too often the unsung heroes of our material culture.

"Our capacity to go beyond the machine rests upon our power to assimilate the machine. Until we have absorbed the lessons of objectivity, impersonality, neutrality, the lesson of the mechanical realm, we cannot go further in our development toward the more richly organic, the more profoundly human."
Lewis Mumford, *Technics and Civilization*, 1934

TASCHEN
A-Z
DESIGNERS AND FIRMS
ROCKET

AEG

FOUNDED 1883
BERLIN, GERMANY

Peter Behrens, Turbine hall for AEG, 1908–1909

→Peter Behrens, Light fitting, 1907

Peter Behrens, AEG advertisement, 1907

In 1881 Emil Rathenau (1838–1915) visited the "Exposition Internationale d'Electricité" in Paris, where he saw **Thomas Alva Edison**'s light bulb. He was so impressed that he bought the patent licence and in 1883 founded the Deutsche Edison Gesellschaft (German Edison Company for Applied Electricity). The company subsequently changed its name to the Allgemeine Elektricitäts Gesellschaft, or AEG. Its catalogue for the 1900 Paris "Exposition Universelle" was designed by the Jugendstil artist Otto Eckmann (1865–1902), who also designed an Art Nouveau-style logo for the company. In 1907, however, AEG appointed as its artistic consultant the architect and designer **Peter Behrens**, who proceeded to create the first wholly integrated corporate identity for the company. Having re-invented AEG's logo, Behrens went on to design not only a unified range of electrical goods such as kettles, clocks and fans, but also the factory buildings required for their production. AEG's production engineer, Michael von Dolivo-Dobrowolsky, realised that the key to the successful mass production of high quality goods lay in the standardization of interchangeable components, which would allow them to be used for several products rather than just one. This type of product standardization and the modern methods of manufacture employed by AEG reflected the ideals of the **Deutscher Werkbund**, of which Behrens had been a founding member in 1907. Hermann Muthesius (1861–1927), another central figure in the formation of the Deutscher Werkbund, had urged German companies to

A. E. G. SPARBOGENLAMPE
ALLGEMEINE ELEKTRICITÄTS-GESELLSCHAFT, BERLIN

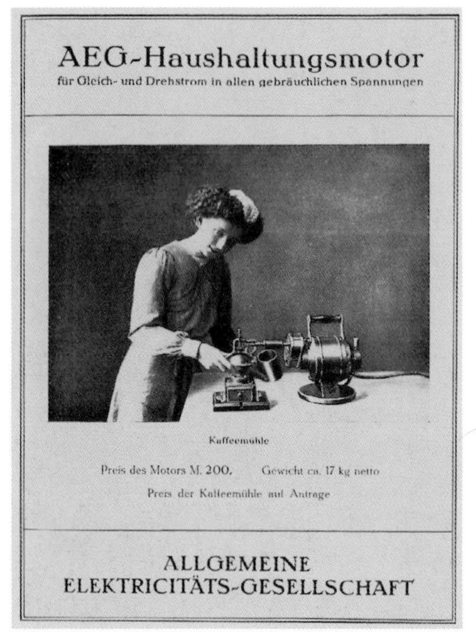

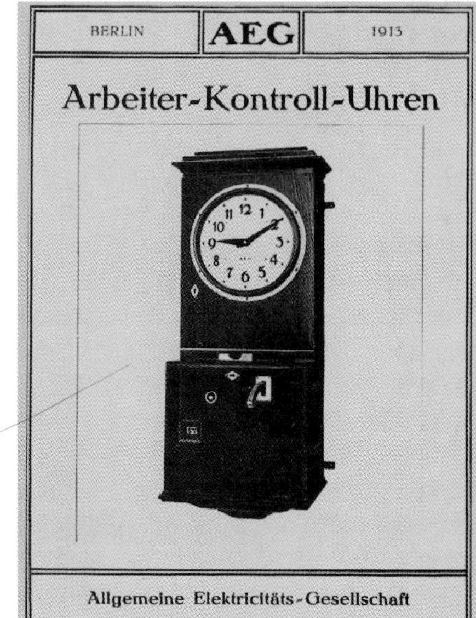

Peter Behrens, Electric coffee grinder, c. 1911

↗Peter Behrens, Clocking-in machine, c. 1910

establish a national aesthetic of "types" and "standards" that would result in "a unification of general taste". AEG was one of the first companies to embrace and implement Muthesius' ideas and to assemble household products from standardized components, as was the case with Peter Behrens' well-known range of kettles and his electric table fan (1908). With a strong brand identity forged by Behrens' comprehensive design programme, AEG went on to become a leading manufacturer of electrical products. In 1927 the company took part in the Deutscher Werkbund's *Die Wohnung* (The Dwelling) exhibition held in Frankfurt, where it displayed electric fans, water heaters, lamps and kettles, among other products. In the 1960s, AEG became primarily known for its almost puristic high quality white goods. In the 1980s, however, AEG entered a financial crisis and was taken over, first by Daimler-Benz, and in 1994 by the Swedish Company Electrolux. Today their concept of high quality and restrained form is very successful and AEG remains a large and prosperous manufacturing company.

EMILIO AMBASZ

BORN 1943
RESISTENCIA, ARGENTINA

Emilio Ambasz studied architecture at Princeton University, taught for a year at the Hochschule für Gestaltung in Ulm and held a professorship at Princeton until 1969. In 1967 he also co-founded the avant-garde Institute of Architecture and Urban Studies in New York. From 1970 to 1976 he was Curator of Design at New York's Museum of Modern Art, where in 1972 he organized the groundbreaking exhibition "Italy: The New Domestic Landscape – Achievements and Problems of Italian Design". He subsequently founded his own design studio, Emilio Ambasz & Associates, in 1977 and the Emilio Ambasz Design Group in 1981 – both in New York. Between 1981 and 1985, Ambasz was president of the Architectural League and taught extensively at Princeton and other American universities. While Ambasz is widely celebrated for his design teaching and writing, he is also acclaimed

Cummins *Signature 600* engine, 1996–1997

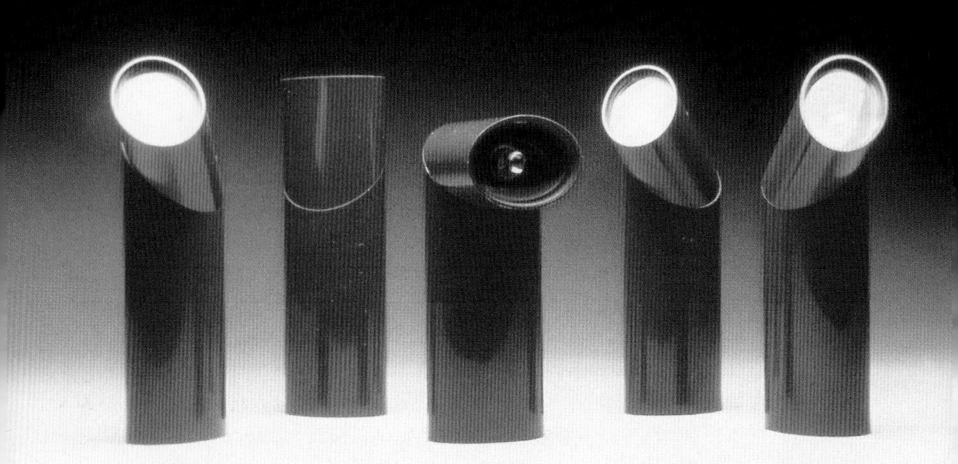

Polyphemus flashlight, 1985

for several notable seating and lighting designs, such as the highly successful seating systems *Vertebra* (1977) with Giancarlo Piretti and *Dorsal* (1981), and the lighting range for Logotec (1981). His architectural projects include the Center for Applied Computer Research and Programming, Las Promesas, Mexico (1975), the Grand Rapids Art Museum, Michigan (1975), the Museum of American Folk Art, New York (1980) and the San Antonio Botanical Garden Conservatory, Texas (1982). Ambasz also won first prize and a gold medal in a competition to design the master plan for Expo 1992 in Seville. Since 1980 Ambasz has been chief design consultant to Cummins Engine Co. and has received numerous awards for his lighting and seating designs – his *Polyphemus* flashlight (1985) was nominated for both a Compasso d'Oro and an IDSA (Industrial Designers Society of America) award in 1987, his *Qualis* seating system (1991) was awarded a Compasso d'Oro in 1991, and his *Soffio* modular lighting system won him the Industrial Design Excellence award by the IDSA. He has stated that the methodological principle guiding his work "is the search for basic principles and prototypical or pilot solutions which can first be formulated into a general method and then applied to solve specific problems". Ambasz believes that design should not just fulfil functional requirements but should also take poetic form in order to satisfy our metaphysical needs. He maintains that designers must learn to reconcile the past and the future in their work and give "poetic form to the pragmatic".

APPLE COMPUTER INC.

FOUNDED 1976
PALO ALTO, USA

Steve Jobs

High school friends Steven Wozniak (b. 1950) and Steven Jobs (b. 1955) shared an interest in electronics but were both perceived as "outsiders" by their peers. After graduation, they dropped out of further education and found employment in Silicon Valley – Wozniak at Hewlett-Packard and Jobs at Atari. In 1976 Wozniak, who had dabbled in the development of computers for some time, designed what would eventually become the *Apple I* computer. Jobs, who was more business-orientated than his friend, insisted that this product should be developed and marketed. As a result, Apple Computers was founded in Palo Alto on 1st April 1976 in the hope that the new personal computer, the *Apple I*, would revolutionize computers just as **Ford**'s *Model T* had revolutionized cars. The venture did not take off until a year later, when the *Apple II* – the first ever commercial personal computer – was launched at a local trade fair. This was the first personal computer to have a plastic casing and to include colour graphics. 1978 saw the introduction of the *Apple Disk II*, which at the time was the cheapest and easiest-to-use floppy drive available. Increasing sales drove the company's rapid growth and by 1980, the year in which the *Apple III* was launched, the business was employing a workforce of several thousand. In 1981 Apple hit the first of several rocky periods in its history; market saturation meant fewer sales, **IBM** released its first PC and Wozniak, the creative force behind the business, was injured in a plane crash and only returned to the company for brief periods of time. Undeterred, Jobs began working on the *Apple Macintosh*, which had initially been conceived as a $500 computer but in the end far exceeded that amount. Its subsequent launch in 1984 marked a real breakthrough in personal computing. With its user-friendly interface, high-definition screen and mouse, the *Mac* heralded the true advent of the home computer. Unlike the *Apple II*, the *Apple Mac* featured an integrated monitor and disk drive. Designed

Apple I, 1976

by Frogdesign, its styling was radically different from its more angular competitors – it had softer, flowing lines that gave it a sleeker aesthetic. Although the smaller, cream-coloured *Mac* initially sold well, users became increasingly frustrated by its small amount of RAM and its lack of hard drive connectivity. Problems also arose through personal differences between Jobs and the then CEO of Apple, John Scully, which eventually resulted in Jobs being ousted in 1985 from the company he had originally created. Bill Gates' introduction of *Windows 1.0*, which had many similarities to the *Mac*'s GUI interface, brought further difficulties. In particular, while Gates had signed a legal statement to the effect that Microsoft would not use *Mac* technology for *Windows 1.0*, it did not extend to future versions of Windows. This resulted in Apple's losing the exclusive rights to its revolutionary, user-friendly icon-based interface. Although the late 1980s were a prosperous time for Apple, by 1990 the market was swamped with cheaper PC clones that could run the newly launched *Windows 3.0* operating system. While watching its market share erode, Apple nevertheless remained the design industry's computer system of choice because of its superior graphics and publishing applications. This was especially true after the launch of the *Powerbook* in 1991 and the *PowerMac* in 1994. By the mid-1990s, Apple was yet again on the brink of collapse with $1 billion of back orders that it could not fulfil because it did not have the parts to make up the products. In 1996, Jobs returned to Apple and began making significant structural changes to the company, after which Apple sold its products directly over the Internet and a cross-licensing agreement was drawn up with its one-time rival Microsoft. In 1998 Jobs oversaw the launch of the company's new landmark product, the *iMac* – another design from Apple that redefined the personal computer. Importantly, the affordable, fun-coloured and translucent *iMac*, the design of which had been overseen by the brilliant young industrial designer **Jonathan Ive**, tapped into the lucrative educational and home markets to the extent that,

Apple logo

Apple Macintosh I
28 K, 1984

by the autumn of 1998, it had become the best-selling computer in America. As Ive notes: "One of the primary objectives for the design of the *iMac* was to create something accessible, understandable, almost familiar." With the success of the *iMac* and its family of products, including the *iBook* and *G3* (1993), the death knell of the anonymous and essentially alienating grey box was finally tolled. Significantly, the *iMac* must also be the first computer to have stylistically influenced the design of other products – from tasking lamps to desk accessories. With its commitment to design excellence and new paradigm innovation, there is little doubt that Apple Computer Inc. will continue to challenge and redefine the personal computer market.

Packaging design for White Rose, c. 1951

EGMONT ARENS

BORN 1888 CLEVELAND, USA
DIED 1966 NEW YORK, USA

Having worked as a sports editor in 1916 for the *Tribune-Citizen* newspaper in Albuquerque, New Mexico, in 1917 Egmont Arens opened his own bookstore in New York. A year later he began printing newspapers under the imprint "Flying Stag Press" and edited the magazines *Creative Art* and *Playboy* (a journal specializing in modern art, not the Hugh Hefner publication of today). Arens later became editor of *Vanity Fair* magazine and also worked at the advertising agency of Earnest Elmo Calkins, where he began his career as an industrial designer by establishing an industrial styling department. Like other well-known industrial designers who gained prominence in America during the 1930s, Arens styled products for manufacturers so as to make them more alluring to consumers, for example the packaging for the grocery chain A&P (his *8 o'clock coffee* package is still in use). He termed this practice "consumer engineering" and wrote extensively on the relationship between design and marketing, most notably in his book *25 Years in a Package*. Arens' many products in the streamline style, including a range of spun aluminium saucepans, were an attempt to "design" America out of the Great Depression.

Egmont Arens & Theodore Brookhart, Meat slicer, *Model No. 410 Streamliner* for Hobart Manufacturing Company, 1941

Poster announcing the change from Horch to Audi, 1910

AUDI

FOUNDED 1909 ZWICKAU, GERMANY
SINCE 1965 INGOLSTADT, GERMANY

In 1896 August Horch (1868–1951) was hired to supervise the mass production of cars at the factory owned by Karl Benz (1844–1929), who had patented his designs for a vehicle powered by an internal combustion engine ten years earlier. After three years with Benz, in 1899 Horch established his own automobile factory in Cologne, Horch & Cie. His first car, designed with a front engine and a new type of gearbox, proved successful and production capabilities rapidly expanded – by 1908 over 100 cars were being manufactured per annum. In 1909, when Horch left the company, Fritz Seidel was appointed its chief designer and Heinrich Paulmann its technical director. Now no longer associated with the company that bore his name, Horch set up another factory in Zwickau in 1909 and employed August Hermann Lange (1867–1922) as his technical director. The venture began producing cars in 1910 bearing the new Audi marque (the latin word for Horch). Audi 4-cylinder cars driven by Horch triumphed at the Austrian Alpine Runs during the

Wanderer W3 automobile, 1913

→Audi *A6 2.7T Quattro*, 1999

↘Audi *A4 1.8T Sport*, 1999

1910s, winning successive Alpine Trophies from 1912 to 1914 and bringing valuable publicity to the new company. Prior to the First World War, Audi had produced five passenger models and two truck models and had a manufacturing capability of 200 vehicles a year. After the war, German society was on the brink of economic collapse. Its currency had gone into free-fall and there was little if any demand for obsolete cars produced with out of date machinery. Between 1925 and 1929, however, German manufacturers began adopting factory-line assembly systems, which allowed the industry to double its productivity. During the 1920s, smaller cars for ordinary people were manufactured. Audi's first post-war model, the *Type K*, was developed by Hermann Lange. In Germany, this was the first car to be offered with left-hand drive and a central gear shift as standard. In 1932, the Audi and Horch companies, together with the vehicle manufacturer DKW and the automotive division of the Wanderer company, joined forces and formed Auto Union AG of Chemnitz – its logo of four rings symbolizing the four entities. The centralization of these companies brought about greater productivity and increased efficiency. Later in 1936, a Central Experimental Department was formed allowing Auto Union to become the first German manufacturer to undertake crash tests. At the end of the Second World War, Auto Union had lost whole factories – its Audi plant was now in the Soviet Military Zone. By 1949, with the help of bank loans, Auto Union re-emerged at a new production site and began producing vehicles, including those with the Audi marque, from its new assembly plant in the Bavarian town of Ingolstadt. By 1965, however, the "Four Rings" had become a wholly owned subsidiary of **Volkswagen**. A year later, a new range of Audis was produced, including the

Audi *920*, 1939

VJS 893
A4
T427 AJO
1.8T

well-equipped Audi *Super 90*. More important models in re-establishing the marque, however, were the Audi *100* (1970), which was developed clandestinely by Ludwig Kraus, and the Audi *80* (1972) by Nuccio Bertone. From 1979, the company decided to move Audi up-market and this resulted in the appearance of the high performance Audi *Quattro* in the spring of 1980. This powerful and solidly built car not only epitomized German engineering excellence, but also revolutionized the international rallying scene. These three Audi models served as design blueprints for subsequent models until the introduction of the Audi *A4*, *A6* and *A8* and sporting *S* versions in 1994. Recently, Audi has developed the light weight, safe and economical *A2* – the world's first volume-production aluminium car. Audi sees itself as defining the future with designs such as these, which are born out of its commitment to "cutting edge technology and visionary design". In 1996 Audi announced its sponsorship of the International Audi Design Award, with the objective of creating a forum for the inter-disciplinary exchange of ideas. According to Audi, "progress is not a matter of abolishing yesterday, but of retaining its essence ... to create a better today". Its famous motto, "Vorsprung durch Technik" (Advancement through Technology) succinctly sums up the company's guiding principles.

Audi *TT Coupé Quattro*, 1999

OSCAR BARNACK

BORN 1879 LYNOW, GERMANY
DIED 1936 BAD NAUHEIM, GERMANY

Oscar Barnack's design for the *Leica I* of 1925 was nothing short of a photographic revolution. Prior to Barnack's designs for the manufacturer Ernst Leitz Wetzlar (later renamed Leica), 35 mm cameras had been relatively bulky and hence difficult to use. Barnack's first 35 mm Leica camera, the *UR-Leica*, was designed and prototyped in 1913 and was eventually put into production around 1918 as the *Leica A*. Rather than using rolls of perforated cine-film, this groundbreaking camera utilized a ridge film cassette, which allowed it to be portable and light. Both the *Ur-Leica* and the later *Leica A* of 1925 (which had a film winder and controls) were highly convenient cameras that had optically superior *Elmar* lenses developed by Max Berck. The *Leica I*, with its distinctive "hockey stick" element on the front of the camera, was launched to great acclaim at the 1925 Leipzig Fair. It set the standard for subsequent Leica designs and ultimately changed the art of still photography. The *Leica I* was designed to be extremely versatile and could be used to photograph anything from microscopic specimens, portraits and landscapes to even the outer limits of the visible solar system. Barnack's designs, which were completely functionally conceived, possessed a very pure quintessentially Machine Age aesthetic.

Leica I camera, 1930

PETER BEHRENS

BORN 1868 HAMBURG, GERMANY
DIED 1940 BERLIN, GERMANY

Peter Behrens studied painting in Hamburg, Düsseldorf and Karlsruhe from 1886 to 1889. In 1890 he married Lilly Krämer and moved to Munich. Initially he worked as a painter, illustrator and calligrapher, and soon after took up arts and crafts. He frequented Bohemian circles in Munich, and was interested in issues relating to the reform of styles of living. His circle of friends included Otto Julius Bierbaum (1865–1910), Richard Dehmel (1863–1910) and Otto Erich Hartleben (1864–1905). In 1892, Behrens the painter co-founded the Munich Secession, whilst his arts and crafts involvement led him to co-establish – with Hermann Obrist (1863–1927), August Endell (1871–1925), Bruno Paul (1874–1968), Richard Riemerschmid (1868–1957) and Bernhard Pankok (1872–1943) – the Vereinigten Werkstätten für Kunst im Handwerk (United Workshops), for serial production of utility objects.
Peter Behrens accepted an invitation from Grand Duke Ernst-Ludwig of Hesse in 1899 to be the second member of the artists' colony then newly established in Darmstadt. There, at the location known as Mathildenhöhe, he built his own house. Self-taught, Behrens designed this as a *Gesamtkunstwerk* (complete-art-work), and from the gardens to the building to the interiors (furniture, lamps, carpets, crockery and glassware, cutlery, towels, décor, paintings etc.) it was created entirely to his designs. The Behrens House marked a significant turning point in the life of the creator. For Peter Behrens it signalled a final farewell to his period in Munich artist circles, and symbolized his relinquishment of Jugendstil and his gravitation toward an austere, sober style of design.
In 1903 he was appointed director of the Düsseldorf College of Arts and Crafts, with a brief to reform it. In this he succeeded brilliantly. Among the new faculty members he brought to Düsseldorf were Josef Bruckmüller, Max Benirschke, Rudolf Bosselt, Fritz Hellmuth Ehmcke and Johannes L. M. Lauweriks.
In 1907 Peter Behrens and ten other individuals – among them Theodor

Kettles illustrated in an AEG catalogue from 1909/1910

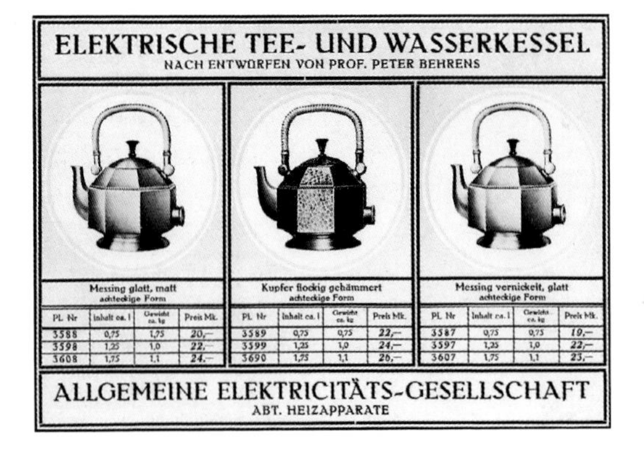

Fischer (1862–1938), Josef Hoffmann (1870–1956), Joseph Maria Olbrich (1867–1908), Bruno Paul, Richard Riemerschmid, and Fritz Schumacher (1869–1949) – as well as twelve companies, including Peter Bruckmann & Sons, the publishing house Eugen Diederichs Verlag, Gebr. Klingsor, and the Vereinigten Werkstätten which Behrens himself had co-founded, joined to found the **Deutscher Werkbund** (DWB). Its establishment was influenced by the Arts & Crafts Movement, but from the outset its aims were of a more modern kind, among them a readiness to embrace industry as a factor in creation, the transformation of a class society into an egalitarian mass society, and the re-humanization of the economy, society and culture. More than virtually any other, Peter Behrens embodied both in his own person and in his work for the **AEG** the ideals of the DWB, which indeed earned him the nickname "Mister Werkbund".

The AEG (Allgemeine Elektricitäts-Gesellschaft, or General Electric Company), founded by Emil Rathenau (1838–1915), had taken on Peter Behrens as its artistic consultant in 1907, on the recommendation of Paul Jordan, who controlled almost all of the AEG plants. It was the first time in the history of industry that a major company had taken a step of this kind, and on such a scale it remains unparalleled to this day. Beginning with minor details (changes to arc lamps), Peter Behrens went on to transform the corporate identity of the AEG, from its letter-head to its logo, the design of its products, its advertising, its exhibition strategy, and indeed many a large factory building. Among these last was his best-known building, the turbine hall flooded with light, hailed as the "cathedral of labour". As artistic consultant, Behrens was neither an employee nor a director of the AEG; he performed his duties as designer and architect on a wholly freelance basis, working in his Neubabelsberg studio near Berlin on commissions from the AEG and other companies and organizations. His other work at this time included the Mannesmann administrative offices in Düsseldorf, the Germany embassy in St. Petersburg, and numerous design jobs,

Electric clock for AEG, 1929

→Electric kettle for AEG, 1909

amongst them the inscription "Dem deutschen Volke" ("For the German people") on the Reichstag building. From 1907 to 1912 the pupils and assistants in his office included Walter Gropius (1883–1969), Ludwig Mies van der Rohe (1886–1969), Charles-Edouard Jeanneret (1887–1965; known later as Le Corbusier), Adolf Meyer (1881–1929), Jean Krämer (1886–1943) and many others who were subsequently to decide the future of modernist design and architecture. Mies van der Rohe admitted to taking his famous maxim "Less is more" from Peter Behrens. "His all-comprehending and fundamental interest in the shaping of the entire environment ... held a great attraction for me," Walter Gropius later observed. The Fagus works built in Alfeld by Gropius and Meyer, like the Bauhaus building in Dessau, continued to explore the interplay of structure and form that the turbine hall had engaged with. Here Behrens set an example for the next generation.
In 1921 Peter Behrens was once again invited to the Düsseldorf College of Arts and Crafts, but in 1922 he accepted a chair at the Academy of Fine Arts in Vienna, where he conducted a master class in architecture. From 1936, following the death of Hans Poelzig (1869–1936), Behrens headed the architecture division of the Prussian Academy of the Arts in Berlin.
During the 1920s, both of Peter Behrens' offices, in Neubabelsberg and Vienna, won commissions for important buildings, such as the administrative offices of the Hoechst paint works (which was to be Behrens' only Expressionist building), the storage depot of the Good Hope Colliery at Oberhausen, a terraced house at Stuttgart's Weissenhof colony, the Villa Gans in Königsstein, the synagogue in Zylina, and the first modernist building in England, the New Ways residence, built in Northampton for Wenman Joseph Basset-Lowke. The complex Behrens built in Linz in the early 1930s for Austria Tabak Regie "is without doubt the finest building constructed in a German-speaking country after 1933" (L. Benevolo). However, the Nazis branded Peter Behrens a "cultural Bolshevik" and excluded him from the award of state commissions.
The achievement of Peter Behrens in the first half of the 20th century was pioneering. His thinking was taken out to the wider world by his pupils, particularly by Gropius, Mies van der Rohe and Le Corbusier. Behrens' invention of the concept of corporate identity had a direct influence on later companies such as **Braun**. The development of a comprehensive philosophy of form is now an accepted maxim. The watchword of Peter Behrens in brief, a message even more relevant today than in his own times, was "the humanization of the industrial world".

(Text: Prof. Till Behrens, Frankfurt/Main)

ALEXANDER GRAHAM BELL

BORN 1847 EDINBURGH, SCOTLAND
DIED 1922 BEIN BHREAGH, NOVA SCOTIA, CANADA

The American audiologist Alexander Graham Bell is best remembered as the inventor of the telephone. He was the third generation of his family, who originated from Edinburgh, to be recognized as a leading authority in elocution and speech therapy. His father's publication, *Standard Elocutionist*, was so popular that it went into nearly 200 editions. Trained by his father, Alexander initially worked as an elocution and music teacher in Elgin, County Moray. In 1864 he became a residential master at the Weston House Academy in Elgin, and it was there that he began his first researches into the nature of sound. Four years later he moved to London, where he became his father's assistant. Following the untimely death of his eldest brother from tuberculosis and the ill health of his younger brother, who had also contracted the disease, in 1870 the surviving Bell family, including Alexander, emigrated to Ontario, Canada. A year later, in Boston, Alexander lectured on and demonstrated his father's system of teaching people with impaired hearing how to speak. In 1872 he founded his own school for training teachers of the deaf, and a year later was appointed professor of vocal physiology at Boston University. While there, Bell began experiments with a young mechanic named Thomas Watson, which resulted in the development of an apparatus that could transmit sound electrically. Bell was granted a patent for a telegraph machine that could send multiple messages, and began outlining the specifications for his new invention, the telephone, which was patented in 1876. That same year, Bell set up the Bell Telephone Company to commercialize his invention. Almost immediately, this attracted the attention of market rivals such as the Western Union Telephone Company, which attempted unsuccessfully to infringe Bell's patent. In France in 1880, Bell was awarded

Alexander Graham Bell speaking for the first time on the newly established telephone connection between Chicago and New York, 18th October 1892

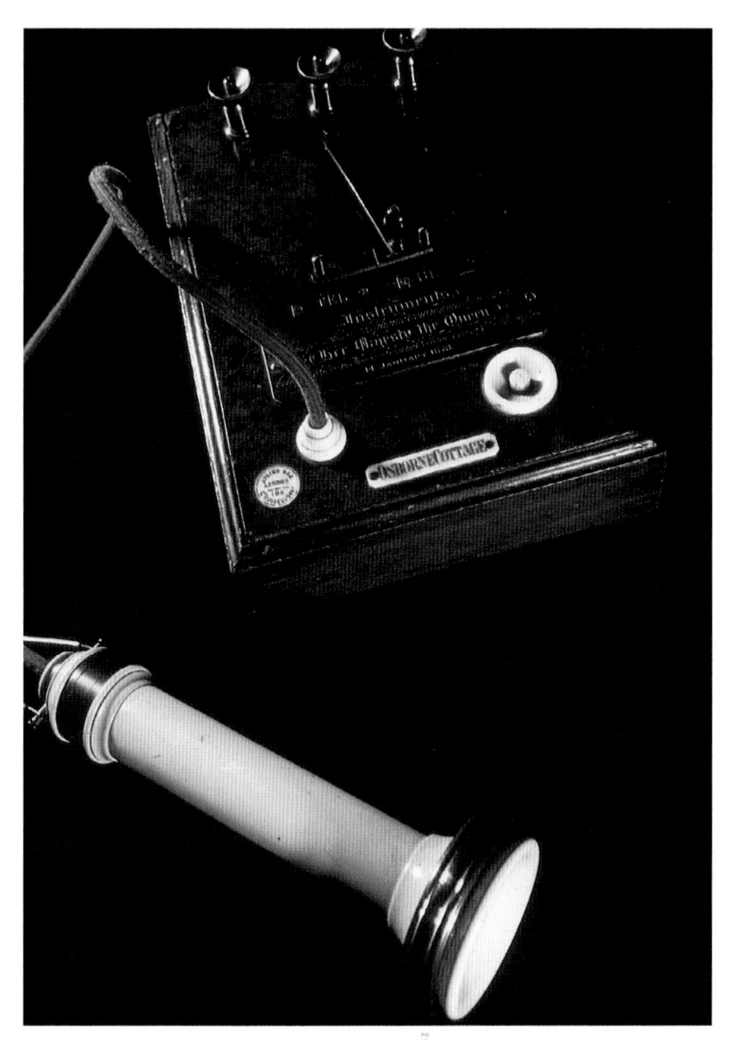

Early Bell telephone and terminal panel, 1877
This example was used by Queen Victoria at Osbourne Cottage in Southampton to communicate with Alexander Graham Bell in London on 14th January 1878.

the Volta prize of 50,000 francs. This enabled him to establish the Volta Laboratory, where with Charles Sumner Tainter and his cousin Chichester Bell, he developed the "Graphophone" in 1886 – a precursor of Emil Berliner's Gramophone. After 1900, Bell experimented with man-carrying kites and undertook research into a number of diverse fields, from sonar detection to hydrofoil watercraft. Bell's revolutionary designs quite literally changed the world of communications and entertainment and in so doing heralded the birth of the Modern Age.

SIGVARD BERNADOTTE

BORN 1907 STOCKHOLM, SWEDEN
DIED 2002 STOCKHOLM, SWEDEN

Facit PI typewriter for Åtvidabergs Industrie, 1958

Margrethe mixing bowl for Rosti, 1954 (co-designed with Acton Bjørn)

The son of King Gustavus VI of Sweden, Count Sigvard Bernadotte (later Prince Sigvard Bernadotte) studied at the University of Uppsala (1926–1929), at the Kungliga Konsthögskolan, Stockholm (1929–1931), and at the Staatsschule für angewandte Kunst, Munich, in 1931, prior to joining the Georg Jensen workshop. He was the first designer there to adopt geometric forms rather than naturalistic forms, as his *Bernadotte* flatware of 1939 demonstrates. During the late 1930s he also designed a volumetric metal and Bakelite cocktail shaker for an English manufacturer. After serving as a director of the Georg Jensen silversmithy, in 1949 he established, a Copenhagen-based design consultancy with the Danish designer Acton Bjørn (1910–1992). This multi-disciplinary studio later expanded, opening branches in Stockholm and New York, and Bernadotte and Bjørn co-designed several notable industrial products, including the melamine *Margrethe* stacking bowls for Rosti (1954) and the compact *Facit PI* typewriter (1958). In 1964 Bernadotte independently founded the Bernadotte Design studio, whose output was distinguished by its use of strong geometric forms and interesting colour combinations. Sigvard Bernadotte is an important pioneer of industrial design in Europe and is also the first European designer to have become a member of the American Designers' Institute.

Biró logo

LÁSZLÓ BIRÓ

BORN 1899 BUDAPEST, HUNGARY
DIED 1985 BUENOS AIRES, ARGENTINA

László Jozsef Biró, a Hungarian-born journalist, created the first ballpoint pen, which he patented in 1938. The principle of the ballpoint pen, however, dates back to 1888, when a patent was filed by the American John J. Loud for a device to mark leather. This design had a number of failings and the patent was not commercially exploited. During the 1930s, while working as an editor of a magazine in Hungary, Biró became aware of an ink used for printing that was quick-drying and smudge-free and so decided to develop a pen that could use the same ink. As this thicker ink would not flow from a regular pen nib, Biró devised a point by fitting his pen with a tiny ball bearing at its tip. As the pen moved along the paper, the ball rotated, picking up ink from the ink cartridge and leaving it on the paper. In 1940 Biró, who was a fervent Communist, left Hungary when it entered into an alliance with Nazi Germany and subsequently emigrated to Argentina. Biró became an Argentinian citizen and adopted a new Spanish-sounding name, Ladislao José Biró. By 1942 his ballpoint pen had been perfected as a result of a special ink formulated by his chemist brother, Georg. After filing a fresh patent in 1943, Biró raised US$ 80,000 worth of backing to allow his pen to go into full-scale mass production. In 1944 Henry George Martin, an English accountant who had invested heavily in Biró's company, Eterpen Co., brought the fully patented ballpoints back to England

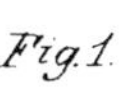

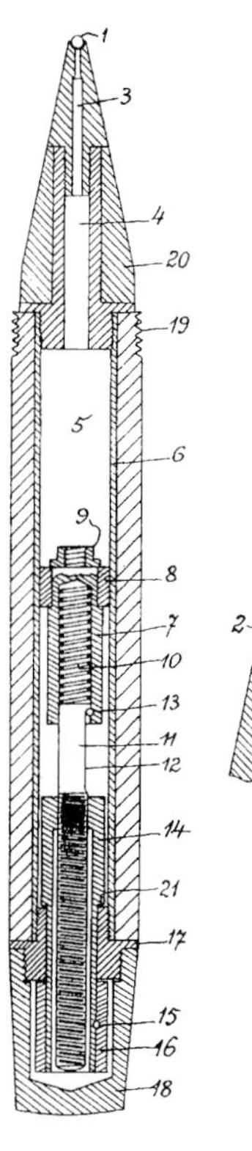

British patent for László Biró's ballpoint pen, 1938

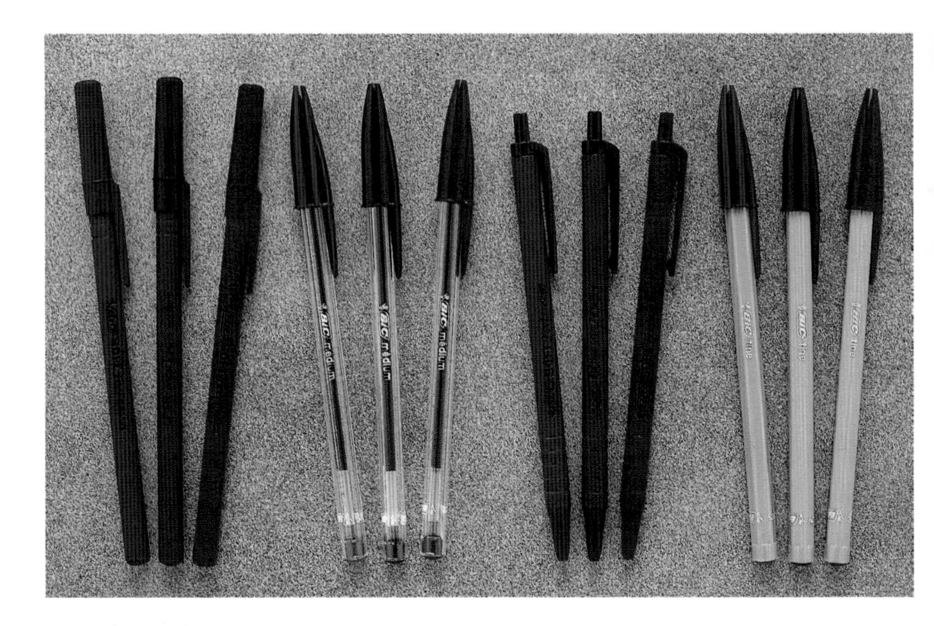

Current range of BIC ballpoint pens, 1999

and offered them to the British and Allied armed forces. Unlike regular fountain pens, Biró's ballpoints were especially useful in unpressurized aircraft because they did not leak. Over 30,000 ballpoints were manufactured for the RAF by the Miles Martin Pen Company in Reading, and an even greater number were produced for the American forces. At the end of the war, the innovative and robust ballpoint pen went on sale to the general public around the world. It was initially seen as a status symbol, no doubt due to its high cost: in America the pen retailed at $12.50. Nevertheless, an incredible 10,000 were sold on its first day of sales. In Britain the pen was priced at 55 shillings (£ 2.74) – the equivalent of a secretary's weekly wage packet. In 1949 Marcel Bich (1914–1994), an expert in plastics machining, perfected his own much less expensive ballpoint pen, which sold under the name "BIC Point", a shortened version of his name. Backed by an innovative advertising campaign, the newly founded Société BIC was selling 42 million units annually after only three years. In 1957, Société BIC purchased Biró Swan, the descendent of the Miles Martin Pen Company. Today, BIC markets a range of some 20 writing instruments that are direct descendants of the original Biró pen. Astonishingly, over 15 million BIC pens are sold every day.

Workmate, the first version was designed by Ron Hickman in 1972

BLACK & DECKER

FOUNDED 1910
BALTIMORE, USA

Black & Decker was founded in Baltimore in 1910 by Duncan Black (1883–1951) and Alonzo Decker (1884–1956) for the manufacture of specialist machinery. Early products included milk-bottle cap machines, letter-graphs, pocket-sized adding machines, candy-dipping machines and machinery for the U. S. Mint. In 1914 the company filed a key patent for a revolutionary electric drill that had a pistol grip and a trigger-style switch. This first ever portable power tool and its later successors brought the company huge success – within a decade of its foundation, Black & Decker's annual sales exceeded US$ 1 million. In 1935 the company commenced manufacturing in the UK. It was not until 1950 that Black & Decker began making tools for home use. Its first domestic product was the famous "Little Red Drill", which had a 1/4" capacity. In 1964 the company developed the *Cordless Zero Torque* tool for NASA's Gemini Space Project. In 1971 Black & Decker designed the *Apollo Moondrill* for the Apollo 15 Mission – a cordless zero-torque tool that could remove core samples from the lunar surface. One of the company's greatest innovations was the *Workmate*, designed by Ron Hickman (b. 1932) in 1972. This multi-use bench was the blueprint for many subsequent variations. In 1979 Black & Decker launched its highly successful hand-held cordless *Dustbuster* vacuum cleaner, and a decade later introduced its *Multi Tools*, including the multiple-use *Multisander*. The development of multi-purpose tools eventually resulted in the launch, in 1998, of the four-in-one *Quattro* tool, which incorporates a jigsaw, drill, screwdriver and sander. Driven as much by state-of-the-art technology as by design innovation, Black & Decker has managed to straddle both the industrial and domestic markets with its extensive range of tools for building, gardening (it launched its first range of lawn-mowers in 1969) and cleaning.

PHH-1 drill, c. 1999

BMW

FOUNDED 1916
MUNICH, GERMANY

BMW *3/15 PS*, 1929–1932

Bayerische Motoren Werke was founded as an aircraft engine manufacturer in 1916. Within just three years Franz Zeno Diemer had set the world altitude record in an aircraft powered by BMW engines. In 1923 the company began producing motorcycles. Five years later, it acquired a car factory in Eisenach and with it, a licence to manufacture a small car, the *Dixi*, which became known as the BMW *3/15 PS*. The first car to be designed in-house was the BMW *3/20*, which was launched in 1932. The company designed several streamlined cars over the course of the 1930s, most notably the BMW *331* and the classic BMW *328* sports car. BMW also dominated car racing during those years, with its aerodynamic *328* roadsters winning their class at the 1938 Mille Miglia. In 1940 the company began mass-producing the *801* aircraft engine – some 30,000 were manufactured over the succeeding five-year period – and also won the Mille Miglia again with a modified "aerodynamic coupé" version of the *328*. In 1941, car production was suspended and resources were directed to the design and construction of rockets and aircraft engines, including one of the world's first jet engines, the *003*. By 1945, however, BMW's factories in Eisenach and Berlin had been lost and the remaining plant in Munich was dismantled. After a three-year production ban, BMW resumed the manufacture of motorcycles in 1948 with the single-cylinder BMW *R24*. In 1951 the company produced its first post-war car, the *501*, but

BMW *331*, 1930s

BMW *Isetta 600*, 1955

BMW *1602–2002* models and the succeeding *3 Series*

BMW *3 Series*, 1998

it was a commercial failure. Four years later BMW launched the diminutive *Isetta* (an italian licence), which bridged the motorbike and automobile markets. This hybrid design was extremely popular in Germany with around 200,000 examples being produced. Having decided to concentrate on the small car market, BMW launched the *Model 700* in 1959 and the *Model 1500* in 1962, which established the design vocabulary of its later compact touring cars. Despite this strategy, the company faced bankruptcy in the late 1960s. Critically, BMW was only able to emerge from the financial doldrums by introducing a range of high-quality, conventionally styled road cars that drove like sports cars. In 1973 the more aggressively styled BMW *2002* became the first mass-produced turbo-powered car in the world. Two years later the *3 Series* was introduced, followed by the *6 Series* coupé and *7 Series* in 1976 and 1977 respectively. During the 1980s and 1990s the company continued to expand; in 1997 it launched its retro-designed *M Roadster*, and a year later introduced the fifth generation of the highly successful *3 Series*. The achievements of BMW have largely rested on the continuity of its brand values, which are based primarily on a commitment to technological innovation and design.

BOEING

FOUNDED 1916
SEATTLE, USA

William E. Boeing

William Boeing (1881–1956) studied engineering at Yale University. In the late 1900s he became fascinated with aircraft and, after a period of some research, concluded that he could build a better bi-plane than any then available. In 1915 he asked the designer, Westervelt, to develop a new aeroplane that he could manufacture. The resulting aircraft was the birch-and-canvas *B&W* twin-float seaplane. During the First World War, Boeing received its first production order from the US Navy for 50 *Model C* seaplanes. Around this time, William Boeing declared: "We are embarked as pioneers upon a new science and industry in which our problems are so new and unusual that it behoves no one to dismiss any novel idea with the statement, It can't be done." Under his careful guidance the company grew and produced a wide range of aircraft – from mail planes and flying boats to military observation planes and torpedo planes. While its competitor, Douglas, was winning international acclaim for its first around-the-world flight in 1924, Boeing was developing state-of-the-art pursuit fighters and was building more fighter aircraft than any other manufacturer. The company's series of fighter aircraft from the 1920s and 1930s included the Boeing *P-26 Peashooter* monoplane, which flew 27 mph faster than any of its bi-plane competitors

Boeing *B52 Stratofortress*, 1948–1952

Boeing *707*, 1955–1957

Boeing *747*, 1963–1968

and employed innovative production techniques in its construction, including the arc-welding of its fuselage frame. With the introduction of anti-trust legislation in 1934, William Boeing grew disheartened and left the firm. The company's new president, Claire Egtvedt, determined that the future lay not just in the development of new military aircraft but also in the design and manufacture of large passenger aircraft, such as the long-range four-engine *Clipper* seaplane. During the Second World War, Boeing massively increased bomber production and by 1944 was building an astonishing 362 *B-17 Flying Fortresses* per month. The enormous *B-17* became legendary for its ability to remain flying even after sustaining severe damage. In total, Boeing plants built 6,981 *B-17*s in various models with another 5,745 being built under a nation-wide collaborative effort by **McDonnell Douglas** and Lockheed. In 1942 Boeing began production of the long-range heavy bomber, the *B-29 Superfortress*, which is mainly remembered as being the aircraft from which the world's first atomic bomb was dropped. During the post-war years Boeing reduced production but did continue developing military aircraft, most notably the *B52 Stratofortress*. Produced between 1952 and 1962, this devastating bomber first saw action in Korea. Remarkably, the last production variation of the aircraft, the *B-52H*, remains in service today – testifying to the superior design and strength of its airframe. During the 1950s the president of Boeing, William Allen, moved the company into the commercial jet market with designs such as the 377 *Stratocruiser* – a luxurious civilian version of the *B-29*. In 1952 the company's net worth was risked so as to begin development of the *Dash-80* commercial jet. The company eventually had to bow to pressure from the airlines and re-design the *Dash-80* so as to match Douglas' *DC-8*. The resulting design was the landmark 707, which featured more powerful engines, a wider fuselage, larger wings and a greater seating capacity than the *DC-8*. The company's family of commercial jets expanded over the years to include the 727, the 737 and the epic 747 jumbo jet. One of the most significant technological achievements in the history of aviation, the 747 is the world's largest commercial aeroplane. Its first flight was in 1969 and it continues to be built at the Boeing factory – the world's largest building by volume in Everett, Washington – clearly demonstrating the aircraft's versatility, popularity, longevity and value. Boeing has also contributed significantly to the American Space Programme, including the famous Apollo missions.

Advertisement for Boeing 707, 1950s

BOSCH

FOUNDED 1866
STUTTGART, GERMANY

Robert Bosch

→Lucian Bernhard, advertisement for the first Bosch spark plug, 1902

Early advertisement for Bosch *Magneto* spark plugs

Robert Bosch (1861–1942) served an apprenticeship as a precision instrument maker with Wilhelm Maier in Ulm from 1876 to 1879, and spent a year in America, working for Sigmund Bergmann and **Thomas Alva Edison**. On his return to Europe in 1885, he stayed briefly in London, where he worked for Siemens Brothers. In November 1886 Bosch opened his own workshop, the Werkstätte für Feinmechanik und Elektrotechnik in Stuttgart, which produced the *Low-Voltage Magneto* for stationary gas engines (1887) and a table-top telephone. By 1898 the company had opened a subsidiary in London, the Compagnie des Magnétos Simms-Bosch, and a year later sales offices were established in Paris and Budapest. In 1901 Robert Bosch's manufacturing operation moved into its first factory, built on the site of the company's existing premises. 1902 saw the launch of the crucial high voltage *Magneto* spark plug, for which the graphic designer Lucian Bernhard (1883–1973) was commissioned to create bold packaging and advertising posters, which helped the company establish its powerful brand identity. In 1909 another factory was built in Feuerbach to increase production capacity. Over the succeeding years, the company diversified its product line by introducing various automotive innovations, including headlamps (1913), the first electric starter motor (1914), car horns (1921), the first electrically powered windscreen wipers (1926) and the first standard diesel injection pump for trucks

At Last You Can Get Them,

THE LONG MISSED AND LONG WANTED

Bosch-Magnetos

For all kinds of Combustion Engines.

AGENCY FOR THE UNITED STATES

of the world-famous Bosch-Magnetos, made by Robert Bosch, Stuttgart, Germany, the inventor of the Bosch Ignition Apparatus, and sole manufacturer of all machines sold as Bosch or Simms-Bosch Magnetos,

HAS NOW BEEN OPENED

Bosch-Ignition is used by all the most important motor car and cycle manufacturers. Some of them are:

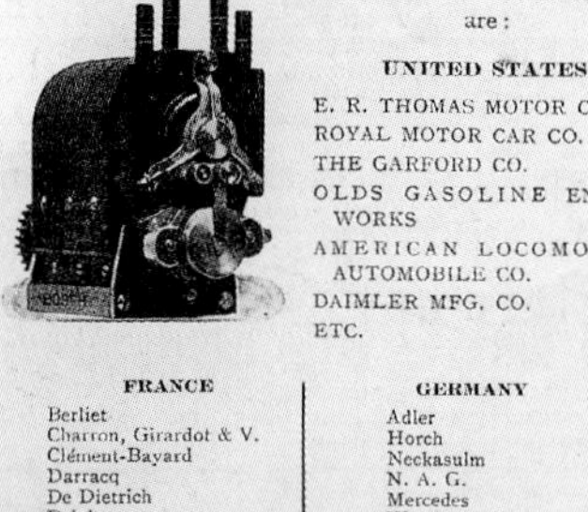

UNITED STATES

E. R. THOMAS MOTOR CO.
ROYAL MOTOR CAR CO.
THE GARFORD CO.
OLDS GASOLINE ENGINE WORKS
AMERICAN LOCOMOTIVE AUTOMOBILE CO.
DAIMLER MFG. CO.
ETC.

FRANCE	GERMANY	GREAT BRITAIN
Berliet	Adler	Daimler
Charron, Girardot & V.	Horch	Napier
Clément-Bayard	Neckasulm	Singer
Darracq	N. A. G.	Westinghouse
De Dietrich	Mercedes	Wolseley
Delahaye	Wanderer	Etc., etc.
Léon Bollée	Etc., etc.	ITALY
Peugeot		Fiat
Renault Frères	BELGIUM	Itala
Richard-Brasier	Fabrique Nationale	Isotta Fraschini
Rochet & Schneider,	Pipe	Rapid
Etc., etc.		Etc., etc.

WRITE, OR BETTER STILL, *CALL* AND OBTAIN PARTICULARS

ROBERT BOSCH, 1947 BROADWAY, New York City

Telephone, 4288 Columbus

BERN
HARD
ROSEN
BOSCH

Advertisement for Blaupunkt in-car radios, 1950s

→*Neuzeit I (New Times)* electric mixer and blender, 1952

→Advertisement for the *Neuzeit I (New Times)* food processor, c. 1952

(1927). In 1932 Bosch took over the Idealwerke AG and produced televisions and radios under the label Blaupunkt. Bosch continued its commitment to research and development throughout the 1930s, introducing the first serial car radio in 1932 and the first standard diesel injection pump for cars in 1936. During that decade the company also began manufacturing white goods and presented its first refrigerator in 1933. Bosch expanded rapidly over the following decades, opening foreign subsidiaries and numerous overseas plants. It also continued to launch innovative products, amongst them the first fuel injection system for cars (1951), the first automatic dishwasher (1964) the first electronic fuel injection system for cars (1967) and the original ABS anti-blocking system (1978). Renowned for its white goods and electric power tools designed by the German industrial designer **Hans Erich Slany** (b. 1926), Bosch maintains its tradition of producing well-designed and well-built products that epitomize the high quality of German industrial design. Today the Bosch Group is one of Germany's largest companies. Its scientists, engineers, designers and technicians are dedicated to the improvement of the function and reliability of existing products and to the development of new products and systems across all areas of its operations, including electrical and electronic automotive equipment, private and public communications technology, power tools, household appliances and thermotechnology as well as automation technology and packaging machinery.

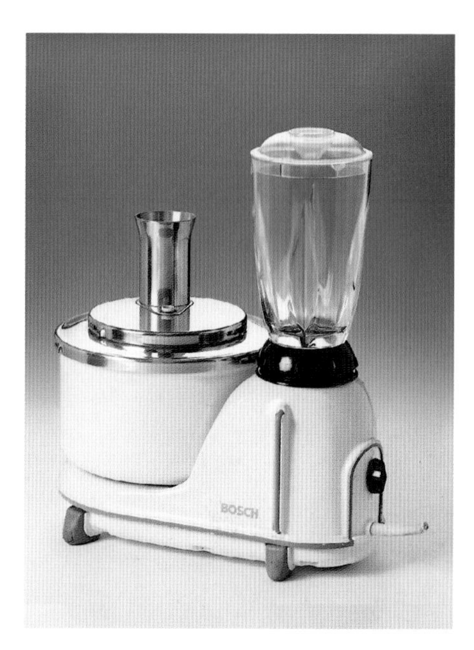

MARIANNE BRANDT

BORN 1893 CHEMNITZ, GERMANY
DIED 1983 KIRCHBERG, GERMANY

Marianne Brandt studied at the School of Fine Art (Großherzoglich-Sächsische Hochschule für Bildende Kunst) in Weimar from 1911 to 1917. She subsequently established her own studio in 1917 and worked as a freelance artist until she enrolled at the Staatliches Bauhaus in Weimar in 1923. After the preliminary course, she took an apprenticeship in the metal workshop, which was then directed by László Moholy-Nagy (1895–1946). Around this time she designed a coffee and tea service, which was based on simple geometric forms – the body of the diminutive teapot being hemispherical. Even at this early date, Brandt appeared more interested in functional form than in traditional craftsmanship and it is not surprising that her later designs became increasingly utilitarian in nature. After taking her journeyman's exam, Brandt became deputy director of the metal workshop in 1928 and organized projects in collaboration with the lighting manufacturers Körting & Mathiesen AG (Kandem) in Leipzig and Schwintzer & Gräff in Berlin. At the Bauhaus she worked alongside fellow metalworkers **Christian Dell** and Hans Przyrembel (1900–1945) and in 1928 co-designed the *Kandem* lamp with Hin Briedendieck (b. 1904), as part of a class project. Brandt's lamps were some of the most important designs to emanate from the Bauhaus because of their suitability for mass production. Although the Bauhaus' focus moved during the 1920s towards the realization of prototypes that were potentially suited to mass-manufacture, very few designs created at the school were successfully put into production. Brandt's lamps were a notable exception. She worked in the architectural office of Walter Gropius (1883–1969) in 1929, and from 1930 to 1933 developed new design concepts for the Ruppelwerk factory in Gotha. In 1933 she returned to Chemnitz, where she took up painting and attempted to license some of her products to the Wohnbedarf department store. Brandt taught at the Staatliche Hochschule für Angewandte Kunst in Dresden from 1949 to 1950 and at the Institut für Angewandte Kunst in Berlin-Weißensee from 1951 to 1954. Although Brandt was an accomplished painter and received recognition for her photomontages, it is her industrial design work that is of greatest significance. Brandt also holds a special position as one of the very first women to design for industrial production – a field that has historically been dominated almost entirely by men.

702 *Kandem* bedside table lamp for Körting & Mathiesen, 1928

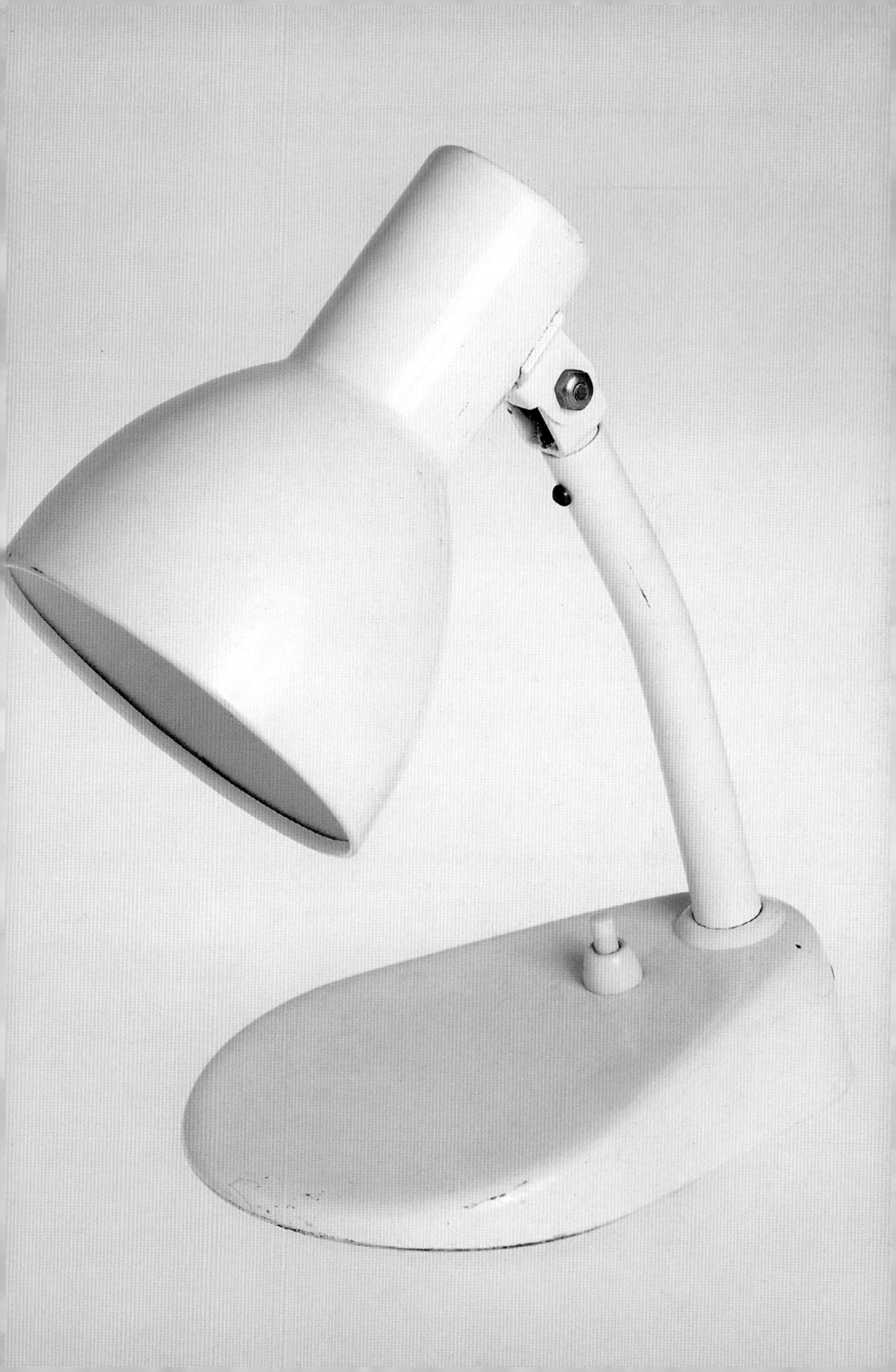

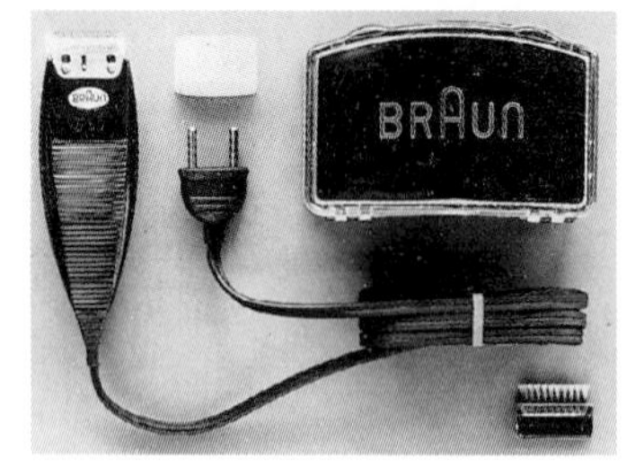

Max Braun, *Model S 50* electric shaver, 1950

BRAUN

FOUNDED 1921
FRANKFURT/MAIN, GERMANY

In 1921 the engineer Max Braun (1895–1946) established a manufacturing company in Frankfurt to produce connectors for drive belts and scientific apparatus. In 1923 he began producing components for the newly emerging radio industry. Following the advent of plastic pellets in 1925, he was quick to seize upon this new material, using homemade presses to manufacture components such as dials and knobs. In 1928 the company moved into a functional modern factory building on Idsteiner Strasse in Frankfurt, and a year later began producing its own radio sets, which were some of the first to incorporate the receiver and speaker in a single unit. In 1932 the company expanded its product range and became one of the first manufacturers to introduce radio/phonograph combination sets. Braun developed a battery-powered radio in 1936 and a year later won an award at the Paris "Exposition Internationale des Arts et Techniques dans la Vie Moderne" for its "exceptional achievements in phonographs".

Max Braun, *Model S 50* electric shaver, 1950

By 1947 the company was mass-producing radio sets, albeit still styled as furniture rather than as Modern electronic equipment. During this period Braun also began production of its *Manulux* flashlights and in 1950 developed its first electric razor, the *S 50*. This shaver incorporated an oscillating cutter-block screened by a thin steel shaver foil – a system that is still used today. In 1950 Braun also branched into domestic appliances with the *Multimix*. After the death of Max Braun in 1951, the firm was headed by his two sons, Artur (b. 1925) and Erwin (b. 1921), who decided to implement a radical design programme that was both rational and systematic. In 1953 Erwin identified a marketing opportunity for distinctive radios that were "honest, unobtrusive and practical devices" and embodied a Modern aesthetic. To this end, Wilhelm Wagenfeld and designers associated with the Hochschule für Gestaltung in Ulm, such as Fritz Eichler (1911–1991), were commissioned in 1954 to redesign the company's radios and phonographs. This new Braun line was introduced at the Düsseldorf Radio Fair in 1955 and attracted international acclaim. 1956 saw the establishment of an in-house design department headed by Eichler, who proceeded to formulate a coherent corporate style based on geometric simplicity, utility and a functionalist approach to the design process. The Braun design vocabulary was not only used for products but was also applied to all areas of corporate identity, including packaging, logos and advertising. Eichler also commissioned other designers associated with the Hochschule für Gestaltung, such as Otl Aicher (1922–1991) and **Hans Gugelot,** to design sleek, unornamented products. Notable designs from this period include Eichler and Artur Braun's line of radios and phonographs (1955) and Dieter Rams (b. 1932) and Hans Gugelot's *Phonosuper SK 4* radio-phonograph (1956), which was nicknamed "Snow White's Coffin". Rams also designed the *Transistor 1* portable radio (1956), the *T 3/T 31* pocket radio (1958) and the first component-based hi-fi system, the *Studio 2* (1959) – all of which helped establish Braun's international reputation. In 1955 **Gerd Alfred Müller** (b. 1932) joined the Braun design team and was responsible for some of the company's best-known designs from the late 1950s, including the *KM 3* multi-purpose kitchen mixer (1957), which embodied the austere rationalist aesthetic that became synonymous with German post-war design. In 1961 Dieter Rams was appointed head of the company's design department, and in 1968 overall director of design. Rams was to head the Braun design team for some 40 years, and his pared-down functionalist aesthetic permeated all the products it manufactured, from kitchen equipment to alarm clocks to electric shavers. During his tenure, Braun introduced a series of landmark designs including the *Per-*

Braun Design Team, hand mixer *MR 300 Compact*, 1987

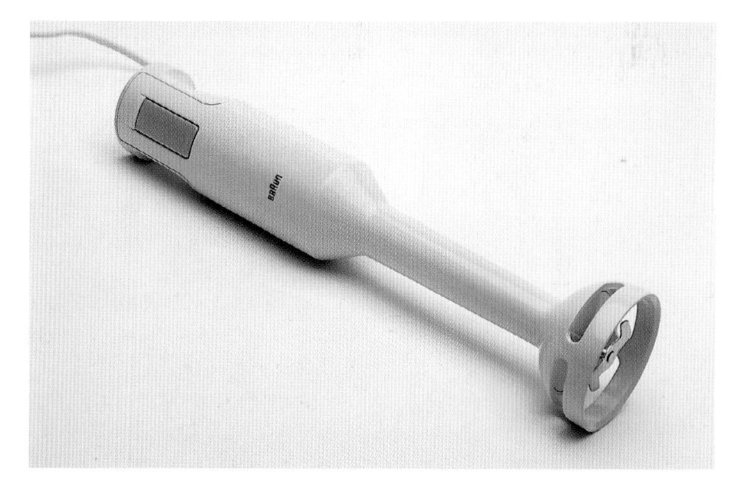

manent lighter (1966), which incorporated an electromagnetic device rather than a traditional friction cylinder, the *ET 22* electronic pocket calculator (1976) and the first radio-controlled clock (1977). In 1967 the Boston-based **Gillette** Company acquired a controlling stake in Braun AG. A year later, the International Braun Awards for design in engineering were established. In 1983 the company was itself awarded the first Corporate Design Award at the Hanover trade fair for its "exemplary conception of product design, information and presentation".

In 1990 Braun discontinued its hi-fi production so as to concentrate on the manufacture of personal grooming products, such as the *Silk-épil EE 1* depilator (1989), the highly successful *Flex Control* line of electric razors (1990) and the *Plak Control D 5* electric toothbrush (1991), as well as a range of hair-dryers. During the 1990s Braun also introduced innovative coffee machines, food processors, hand mixers, irons and alarm clocks. In 1996 Braun launched the *Thermoscan* infrared thermometer, which marked its entry into the personal diagnostic appliance market. Certainly, Braun's success stems from the fact that its products are jointly developed by designers, engineers and marketing experts in accordance with basic design principles. The company uses design innovation to achieve technical and functional innovation and has established a tradition of progressiveness within its

Ludwig Littmann, *K 750 Combi Max* food processor, 1997

design team. The strong aesthetic clarity of its products is the outcome of a logical ordering of elements and the quest for a harmonious and unobtrusive totality. Braun acknowledges that "integrated working methods are ultimately reflected in the obviousness of the product expression", and asserts that "Braun Design is the orientation towards lasting worthwhile values: innovation, distinctive, desirable, functional, clear, honest, aesthetic."

Roland Ullmann, *Flex Integral Colour Selection* razors, 1997–1998

Faxphone 8, 1988

CANON

FOUNDED 1947
TOKYO, JAPAN

The Precision Optical Instruments Laboratory was established in Tokyo in 1933 to conduct research into camera technology. A year later, the company developed a prototype of the *Kwanon*, Japan's first 35mm focal-plane shutter camera, and in 1940 designed Japan's first indirect X-ray camera. Introduced in 1946, the Canon *SII* camera was acclaimed by officers from the Occupation Forces and also by foreign purchasers. In 1947 the firm was renamed the Canon Camera Company and over the following years continued to produce innovations in camera technology, such as the world's first speed-light synchronized flash and shutter camera, the Canon *IVSb* (1952). The Canon *L1* still-camera and the Canon *8T* cine-camera became the first products to receive a "Good Design" designation from the Japanese Ministry of International Trade and Industry in 1957. During the 1960s and 1970s, Canon diversified its product line to include microfilm systems and calculators. In 1965 the company entered the copying machine industry with its introduction of the *Canofax 1000*. Three years later it also launched the first-ever four-track, four-channel recording head. Throughout the 1970s Canon continued developing

Kwanon camera – Japan's first 35mm focal-plane shutter camera, 1934

state-of-the-art cameras, including its top-end SLR, the Canon *F-1* (1971) and the landmark *AE-1* with a built-in microcomputer (1976), which triggered an *AE* SLR boom. In 1975 the company succeeded in developing a laser beam printer (LBP) and in 1976 introduced the world's first non-mydriatic retinal camera, the *CR-45NM*. This was followed in 1981 by the first bubble-jet printer and the new *F-1* SLR. In 1982 Canon launched a colour ink-jet printer and in 1986 the first multi-functional telephone with a built-in fax. Around this time, Canon also began producing digital camcorders, the *Optura* (1997) and the lighter and more compact *Elura* (1999). Over the company's 56-year history, the cumulative production of its cameras has exceeded 100 million units. From handshake-compensating binoculars to digital cameras to ultra-high-quality digital imaging equipment and high-definition television broadcasting lenses, all Canon products have a strong brand identity. Among the company's product design guidelines are: the use of advanced forms, materials and technology; the optimization of comfort with regard to human factors; the dedication to ease of operation; the consideration of the operational environment; and the creation of transcultural solutions.

G-Mark image stabilizer binoculars, 1995

CHRYSLER

FOUNDED 1925
DELAWARE, USA

Walter Percy Chrysler with a model of Carl Breer's *Airflow*, 1930s

Walter Percy Chrysler (1875–1940) began his working life as a machinist's apprentice in the railroad industry, but by the age of 33 had already become the superintendent of motive power for the Chicago Great Western Railway. In 1910 he moved to the American Locomotive Company to manage its Pittsburgh works and around this time purchased his first car, a Locomobile Phaeton. A couple of years later, Chrysler joined the Buick Motor Car Company in Flint, Michigan. In 1916 Buick became **General Motors**' first automotive division, and in 1917 Chrysler was appointed its president and general manager. He was promoted to vice-president of General Motors in 1919 but retired from the company a year later. He next worked for Willys-Overland and then the ailing Maxwell Motor Car Company, which he revitalized with the development of the Chrysler *Six* (1924) – America's first high-styled, medium-priced automobile, which set an industry sales record by selling 32,000 units. The Chrysler Corporation was established in Delaware in 1925 as the successor to the Maxwell Motor Car Company and Walter Chrysler was named its chairman. The same year saw the introduction of the highly successful Chrysler *Four*, with a top-speed of 58 mph, and by 1926 the corporation had risen from 57th to 5th place in industry sales. In 1928

Walter Percy Chrysler with *Four Series 58*, 1925

Dodge truck, 1940s

Chrysler manufactured the first Plymouth and De Soto models and the company also acquired Dodge Brothers, which was highly regarded for its utility vehicles. Although hit by the Great Depression, Chrysler did not cut back on its research and development programme and pioneered several design and engineering innovations, including the "Floating Power" engine mounting system, which reduced vibration to produce a smoother ride. Although the company hired its first in-house stylist, Ray Dietrich (1894–1980), in 1932, design and styling at Chrysler essentially took a back seat to engineering until 1950. A notable exception was the Chrysler *Airflow* of 1934. Designed by the company's chief engineer, Carl Breer (1883–1970), and his staff engineers Fred Zeder and Owen Skelton, the *Airflow* introduced leading-edge aerodynamic styling, innovative weight distribution, unitized body construction and the industry's first one-piece curved windscreen. While the *Airflow* was way ahead of its time and consequently a commercial failure, it was immensely influential upon later automotive design.
During the Second World War, Chrysler produced 18,000 32-ton *M-4 Sherman* tanks and around 500,000 Dodge trucks for military use. By 1945 the company had supplied over US$3.4 billion worth of equipment to the Allied Forces. After the war, Chrysler responded to the need for increased car and truck production by building or buying 11 plants between 1947 and 1950. In 1949 the company also appointed its first director of advanced styling, Virgil Max Exner (1909–1973), who had previously worked in **Raymond Loewy**'s

Dodge *Viper*, 1992

design office. Exner introduced European-like styling to the company and developed the "Forward Look", a style that emphasized movement and speed through the use of curved side windows and tail fins. The Forward Look is perhaps best exemplified by Exner's 1957 Plymouth *Belvedere*. During the 1950s Chrysler also introduced several innovations including power steering, key-operated ignition, electric windows and cushioned dashboards for improved safety. In 1951 it also developed the legendary "Hemi" V-8 engine. With deep hemispherical combustion chambers, the Hemi featured large valves that provided high-volume efficiency and produced 20 % more horsepower than standard V-8 engines. The Hemi was discontinued in the mid-1950s but reappeared in 1964 as the ferocious 425 horsepower 426 Hemi.

Chrysler 392 Hemi *Firepower V-8* engine, 1951

While it was in production, the 426 Hemi was the perfect power plant for Chrysler's emerging "muscle" cars. Elwood P. Engel (1917–1986) was the designer chiefly responsible for the creation of such beefy and brutishly purposeful cars as the Dodge *Challenger*, the Dodge *Charger*, the Plymouth *Road Runner* and the Plymouth *Barracuda*. From the mid-1960s to the early 1970s, these cars dominated the muscle-car street scene and became

Chrysler *Voyager*,
1999 model

legends in their own time. The oil crisis in the early 1970s put an end to this period of excess in the American car industry, however, and prompted a demand for smaller, more fuel-efficient vehicles. Chrysler found itself facing a financial crisis, and in 1979 the celebrated auto executive Lee Iacocca was brought in to turn the company around. With the well-publicized challenge, "If you can find a better car, buy it", Iacocca managed to save the company with the introduction of the "K-cars" in 1981 – the Dodge *Aries* and the Plymouth *Reliant*. This was followed in 1983 by Chrysler's introduction of the minivan in the form of the Dodge *Caravan* and Plymouth *Voyager*. These vehicles not only created a whole new market – people carriers – but also became the company's best sellers. In 1987 Chrysler re-entered the European market with export vehicles and also purchased the exotic Italian sports car manufacturer, Lamborghini. During the 1990s Chrysler launched several notable and financially successful models, including the awesome 700 hp Dodge *Viper* (1992), the Jeep *Grand Cherokee* (1993) sports-utility vehicle and a striking evolution of the *Voyager* MPV (1995). By adapting vehicle design to meet changes in living patterns and customer expectations, Chrysler has managed to survive and prosper. In 1998 the company merged with the prestigious German automotive manufacturer, Daimler-Benz AG, and its product range has consequently adapted to the European market.

CITROËN

FOUNDED 1913
PARIS, FRANCE

Flaminio Bertoni, Citroën *DS19*, 1962

→Pierre Boulanger, prototype of the Citroën *2CV*, 1936

↘Citroën *2CV CI*, 1959

André-Gustave Citroën (c. 1878–1935) trained at the École Polytechnique and then worked as an engineer and industrial designer. From 1905 he began manufacturing components for cars and introduced double helical gears to France. In 1913 he founded the Société des Engrenages Citroën in Paris and two years later began manufacturing ordinance – up to 55,000 shells per day – for the French army. At the end of the war, with arms sales falling, Citroën converted his factory to the production of small and inexpensive cars and hired Jules Saloman as his first automotive designer. In 1919 the company launched the first mass-produced car in Europe, the *Type A*. This car significantly advanced the standards of automotive design with its disc wheels, electric lights and on-board starter. It was, however, the slightly later, more compact and much better-selling *5CV* that firmly established Citroën's reputation in the mass market. In 1934 André Citroën almost bankrupted his business by over-investing in tooling for the *Traction Avant* (later launched as the *7A*) and the company was then taken over by Michelin. The *Traction Avant* was an advanced prototypical design that was at least two decades ahead of its time, with front wheel drive and a long wheel base. Citroën's best-ever seller, the *2CV*, was designed by Pierre Boulanger (1886–1950) and first appeared in 1939. Launched in 1948 as a functionalist, no-frills work-horse, the *2CV* was intended to rival **Volkswagen**'s *Beetle*. Between 1948 and 1990, when production ceased, nearly 3,870,000 *2CVs* were built. Citroën's famous and more elegant *DS19*, designed by Flaminio Bertoni (1903–1964) in 1955, was the evolutionary successor of the *Traction Avant*. With its low-slung forward-looking aesthetic, the *DS19* became known as the "Goddess" in France and was, like the *Traction Avant*, an extremely advanced vehicle for its time. Citroën was eventually taken over by Peugeot in 1975, and although it continues to produce very worthy vehicles, the level of design innovation it achieved in the past has yet to be matched.

Citroën poster showing the *Type A*, 1919

Christopher Dresser, Coalbrookdale Co. cast-iron grate, 1879

COALBROOKDALE

BIRTHPLACE OF THE INDUSTRIAL REVOLUTION
SHROPSHIRE, ENGLAND

Abraham Darby (c. 1678–1717), a British ironmaster, pioneered the smelting of iron ore with coke in the Coalbrookdale area of Shropshire during the early 18th century. Coalbrookdale's close proximity to supplies of natural resources, low-sulphur coal in particular, helped revolutionize the production of iron. In 1709 Darby was the first to produce marketable iron in a coke-fired furnace and went on to manufacture the first iron rails, iron boat and steam locomotive in Britain. Such was the quality of Darby's iron that it could be cast in thin sheets and could thus compete successfully with brass in the manufacture of pots and other hollow ware. The famous cast-iron bridge near Coalbrookdale was also one of the first of its kind. Designed by Thomas Pritchard, the bridge spans 43 metres across the River Severn and was erected between 1777 and 1779 by John Wilkinson and Darby's grandson, Abraham Darby III (1750–1791), partly to advertise the family iron foundry. As a major centre of iron production for over 100 years, Coalbrookdale was renowned in the 19th century for its cast-iron fireplaces and hall furniture, some designed by Christopher Dresser (1834–1904). As the birthplace of the Industrial Revolution, Coalbrookdale is now a world heritage site.

Philippe Jacques de Loutherbourg, *Coalbrookdale by Night*, 1801

WELLS COATES

BORN 1895 TOKYO, JAPAN
DIED 1958 VANCOUVER, CANADA

Wells Coates studied engineering at the University of British Columbia, Vancouver, and graduated in 1921. He subsequently moved to Britain and from 1922 to 1924 studied for a doctorate in engineering at London University. From 1923 to 1926 he worked as a journalist for the *Daily Express* and for a brief period was one of their correspondents in Paris. During this time he wrote from a humanist perspective, viewing design as a catalyst for social change. In London in 1928, he designed fabrics for the Crysede Textile Company and interiors for the firm's factory in Welwyn Garden City, in which he used plywood elements. From 1931 Coates was a consultant to Jack Pritchard's plywood products company, Isokon – a firm pioneering Modernism in Britain. Coates was also commissioned by Pritchard in 1931 to design the Lawn Road Flats, Hampstead, which are seminal examples of British Modern Movement architecture. In 1933 Coates co-founded MARS (Modern Architecture Research Group) and established design partnerships with Patrick Gwynne in 1932 and David Pleydell-Bouverie in 1933. From 1932 he designed a series of Bakelite radios for the Ekco Radio Company, including his famous circular Ekco *AD 65* (1934), which were conceived for industrial production and were among the first Modern products available to British consumers. After the Second World War Coates worked in Vancouver, designing interiors for de Havilland and BOAC aircraft, and in the 1950s producing designs for television cabinets. Just as Wells Coates believed that "the social characteristics of the age determine its art", so his designs epitomize the flowering of British Modernism in the 1930s.

Ekco *AD65* radio for E. K. Cole, 1934

MICHELE DE LUCCHI

BORN 1951
FERRARA, ITALY

Michele De Lucchi studied at the Liceo Scientifico Enrico Fermi in Padua and later trained as an architect under Adolfo Natalini (b. 1941) at Florence University, graduating in 1975. In 1973, together with Piero Brombini, Pier Paola Bortolami, Boris Pastrovicchio and Valerio Tridenti, he founded the architecture and design group Cavart, which promoted a radical design agenda through happenings, publications and seminars. He also collaborated with Superstudio, Ettore Sottsass (b. 1917) and Gaetano Pesce (b. 1939) and taught architecture at Florence University from 1975 to 1977. He then moved to Milan and began working as a design consultant to the **Kartell** in-house design studio, Centrokappa. De Lucchi later assisted Ettore Sottsass with the planning of the first Memphis exhibition. In 1979 he designed several Post-Modern prototypes of domestic electrical appliances for Studio Alchimia and became a consultant to **Olivetti**. He co-founded Memphis in 1981 and was responsible for the introduction of geometric motifs on the plastic laminates used by the co-operative. In 1986 he founded Solid, a Milan-based design group, and also began teaching at the Domus Academy, Milan. In the early 1990s he established the De Lucchi Group and has since worked widely in Japan and Germany. Increasingly, his designs have become more suited to industrial production and his clients have included Artemide, Kartell, Bieffeplast, Mandarina Duck and Pelikan. He has won numerous design prizes, including Good Design awards (Japan), Die Gute Form and Deutsche Auswahl awards (Germany) and a Compasso d'Oro (Italy). De Lucchi regards design as a means of communication, a sentiment that has informed his work throughout his career – from contesting young radical to established international industrial designer.

Tolomeo wall light for Artemide, 1995

JOHN DEERE

BORN 1804 RUTLAND, VERMONT, USA
DIED 1886 MOLINE, ILLINOIS, USA

John Deere served a four-year apprenticeship as a blacksmith in Vermont, before working as a journeyman. He soon gained considerable fame for his careful workmanship and ingenuity. His highly polished hay forks and shovels were in especially high demand throughout Western Vermont. Tempted by tales of a better life in the West, in 1836 Deere moved to Grand Detour, Illinois, where he entered into a business partnership with Major Leonard Andrus. Deere quickly discovered that the cast-iron ploughs originally intended for the light sandy soils of New England could not cope with the clod-like earth of the Mid-West that the pioneers were trying to farm. Having to frequently repair their damaged cast-iron ploughs, Deere became convinced that the problem could be overcome by the addition of a highly polished mouldboard that, if shaped properly, would scour itself clean while turning the furrow slice. In 1837 Deere designed the first commercially successful self-scouring steel plough and began manufacturing the "self-polishers" to sell to local farms. This landmark design, which transformed farming in the West, was followed by another 50 improved models over the next two years. In 1843 Deere managed to arrange the importation of steel from England, and by 1846 his business was manufacturing around 1,000 ploughs per annum. He subsequently established a factory in Moline, Illinois, which by 1857 was producing 10,000 ploughs a year. Deere constantly changed and

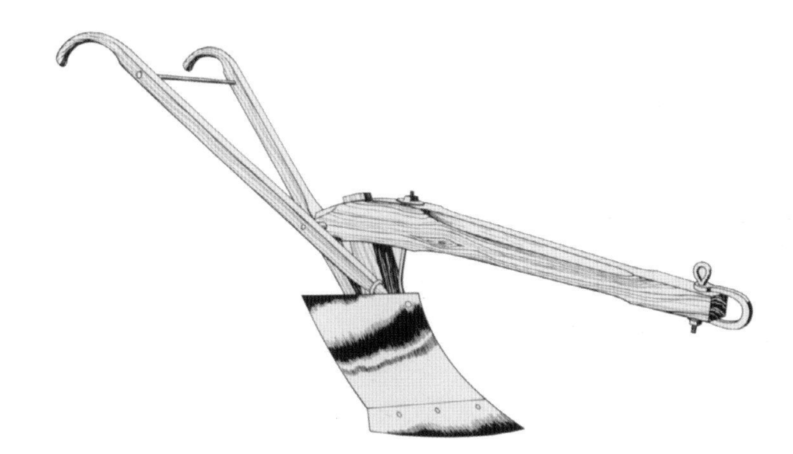

Self-scouring steel plough, 1837

improved the design of his ploughs, declaring: "If we don't improve our product, somebody else will." The company was incorporated as Deere & Company in 1868 and the following year John Deere's son, Charles, succeeded him as president. In 1918 the company purchased the Waterloo Gasoline Traction Engine Company in Waterloo, Iowa, and tractors became an important part of the John Deere line. Deere & Company introduced its first tractor, the *Model D*, in 1923, and over the succeeding decades the company diversified its product-line to include cultivators, harvesters, balers and other engine-powered agricultural machinery, as well as construction equipment such as backhoes and loaders. Deere & Company is now the largest farm equipment manufacturer in North America.

↑*D* tractor, 1923, the first Waterloo tractor to bear the name of John Deere (produced 1924 to 1953)

GP tractor, 1928 (produced until 1935)

MR tractor, 1992

CHRISTIAN DELL

BORN 1893 HANAU, GERMANY
DIED 1974 WIESBADEN, GERMANY

During his apprenticeship as a silversmith at Schleißner & Söhne in Hanau, Christian Dell also studied at the Königliche Preußische Zeichenakademie. From 1911 to 1912 he worked as a silversmith in Dresden before training under Henry van de Velde (1863–1957) at the Kunstgewerbeschule in Weimar from 1912 to 1913. After the First World War he worked as a journeyman and later as a master silversmith for Hestermann & Ernst in Munich. In 1920 he joined the silver workshop of Emil Lettré (1876–1954) in Berlin, establishing his own silver studio a year later in Hanau. From 1922 to 1925 he headed the metal workshop at the Weimar Bauhaus, and then taught for seven years at the Frankfurter Kunstschule, where he designed silverware that was made in the school's workshop. During this period, Dell began producing designs that were suitable for industrial production. His lights, such as the adjustable *Rondella* desk lamp (1927–1928) and his *Idell* range, which was later copied by Helo, were mass-produced by Rondella and Kaiser respectively. Dell designed over 500 lights in total and his *Idell* range remained in production for over 60 years. In the 1930s Dell began experimenting with plastics and in 1939 established his own jewellery business in Wiesbaden.

↘*Das Frankfurter Register 1* brochure, 1928

Rondella desk lamp for Rondella, 1927–1928

DAS FRANKFURTER REGISTER 1
„RONDELLA" TISCH- UND STÄNDER- (ATELIER) LAMPE

DIE STÄNDER-(ATELIER) LAMPE

HÖHE 1,78 cm PREIS: 75 M.

DIE TISCHLAMPE

PREIS: Bessere Ausführ. 33 M. Einfache Ausführ. 25 M.

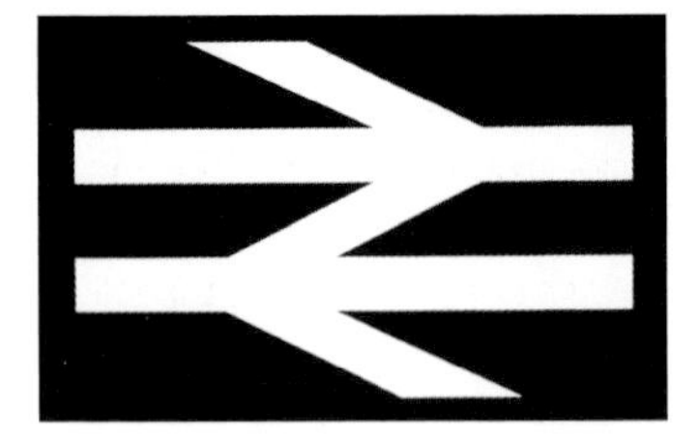

Corporate identity for British Rail, 1965

DESIGN RESEARCH UNIT

FOUNDED 1943
LONDON, ENGLAND

In 1922 Milner Gray and Charles and Henry Bassett founded the Bassett-Gray Group of Artists and Writers, one of the first professional design practices in Britain. Collaborating with artists such as Graham Sutherland (1903–1980), Bassett-Gray represented "a body of artists who design for industrial and commercial purposes". In 1934 Bassett-Gray was joined by Misha Black. The following year it was reorganized and renamed the Industrial Design Partnership, even though at this time the practice was still primarily focused on graphic, packaging and exhibition design. Following the outbreak of the Second World War, the IDP was shut down and both Gray and Black joined the Ministry of Information. Realizing that there would be a huge demand for consumer goods once wartime restrictions had been lifted, in 1942 Marcus Brumwell (1901–1983) and the design theorist Herbert Read (1893–1968) proposed the idea of forming a collective

Exhibition sales caravan for Olivetti, c. 1956

of designers and architects who could produce designs for large-scale industrial production after the war. Milner Gray, who was by then head of exhibition design at the Ministry of Information, was consulted on the organization of such a group. The Design Research Unit was subsequently established in January 1943 and from the outset emphasized industrial product design. Both Black and Gray joined the DRU, the former originating the majority of its product designs – from televisions and electric heaters to cameras and trains – while Gray concentrated chiefly on packaging and signage. Dorothy Goslett, who helped administrate the office, published a book in 1961 entitled *Professional Practice for Designers*, which was in essence a guide to running a design consultancy. "Designers," she wrote, "have to deal directly with business men who are seldom likely to be sentimental about creative work. ... The relationship between designer and client is therefore strictly a business one. The designer's codes of professional conduct are the foundations on which this relationship is built." As the highest-profile design office in post-war Britain, the DRU did much to elevate the professionalism of industrial design practice and provided "an effective design service for industry" by giving concrete expression to Modern design theory.

Ilford *Pixie* camera, c. 1960s

DONALD DESKEY

BORN 1894 BLUE EARTH, MINNESOTA, USA
DIED 1989 VERO BEACH, FLORIDA, USA

Packaging design for Procter & Gamble, c. 1951

Donald Deskey studied architecture at the University of California, Berkeley, as well as fine art at the Arts' Student League, New York, and the Institute of Chicago's School of Art. He later trained in Paris at the École de la Grande Chaumière, the Académie Colarossi and the Atelier Léger. After the First World War he worked as a graphic designer for a Chicago advertising agency, but a visit to the 1925 Paris "Exposition Internationale des Arts Décoratifs et Industriels" prompted him to turn his attention to three-dimensional design. He subsequently produced a number of screens for Saks Fifth Avenue in 1926 and a year later designed window displays for this New York department store as well as for Franklin Simon. Also in 1927, he established the Deskey-Vollmer partnership with Philip Vollmer, which concentrated on exclusive metal furniture and lighting, but this venture was dissolved in the early 1930s. During the late 1920s, he developed a stained-wood laminate known as Weldtex, while in 1931 he designed Moderne interiors for the Radio City Music Hall – his most prestigious and influential commission, which exemplified the American Art Deco style. During the 1930s he executed a variety of industrial designs, such as washing machines, vending machines and even a rack for bowling balls, which were exhibited at the Metropolitan Museum of Art in 1934. From the late 1930s until 1975, he was the principal of Donald Deskey Associates and designed everything from graphics and packaging for Procter & Gamble to printing presses for American Type Founders. His design work was characterized by streamlined forms and innovative planning. His sleekly styled filing cabinet for the Globe-Wernicke Company, for instance, had several novel features including a combined drawer pull/identification panel and drawer fronts with contoured edges. While chiefly remembered as one of the greatest proponents of the Art Deco style in America, Deskey was also an important pioneer of industrial design consulting.

Filing cabinet for Globe-Wernicke, c. 1951

Deutscher Werkbund emblem

DEUTSCHER WERKBUND

FOUNDED 1907
MUNICH, GERMANY

Founded in October 1907, the Deutscher Werkbund attempted from its outset to reconcile artistic endeavour with industrial mass production. Its founding members not only included designers such as Richard Riemerschmid (1868–1957), Bruno Paul (1874–1968), **Peter Behrens** and Josef Maria Olbrich (1867–1908), but also a dozen established manufacturers, including Peter Bruckmann & Söhne and Poeschel & Trepte, and design workshops such as the Wiener Werkstätte and the Munich-based Vereinigte Werkstätten für Kunst im Handwerk. Peter Bruckmann (1865–1927) was appointed the association's first president and within a year its membership had risen to around 500. From 1912 the Werkbund began publishing its own yearbook, which included illustrations of and articles on its members' designs, such as factories by Walter Gropius (1883–1969) and Peter Behrens and cars by Ernst Naumann. The yearbook also listed members' addresses and areas of specialization in an attempt to promote collaboration between art and industry. In 1914 the Werkbund organized a landmark exhibition in Cologne, which included within its grounds Walter Gropius' steel and glass model factory, Bruno Taut's Glass Pavilion and Henry van de Velde's Werkbund Theatre. A year later, the Werkbund's membership had swollen to almost 2,000. The increasing divergence between handcraftsmanship and industrial production fuelled heated debate within the Werkbund, however, with members such as Hermann Muthesius (1861–1927) and Naumann urging for standardization, while others such as van de Velde, Gropius and Taut argued for individualism. This conflict almost led to the disbanding of the association. The widespread need for consumer products after the devastation of the First World War, however, led Gropius to accept the necessity for

Fritz Hellmut Ehmcke, tobacco box for the Deutscher Werkbund exhibition in Cologne, 1912–1914

AEG stand in the main hall of the 1927 Deutscher Werkbund "Die Wohnung" exhibition in Stuttgart

standardization and industrial production, although other members, such as Hans Poelzig (1869–1939), continued to resist change. From 1921 to 1926 Riemerschmid was president of the Deutscher Werkbund and during his tenure the functionalists' approach to design was advanced. In 1924 the Werkbund organized the exhibition *Form ohne Ornament* (*Form without Ornament*). The catalogue illustrated industrially-produced designs and promoted throughout its text the virtues of plain undecorated surfaces and, ultimately, functionalism. In 1927 the Werkbund staged a unique exhibition in Stuttgart, entitled "Die Wohnung" (The Dwelling), which was organized by Ludwig Mies van der Rohe (1886–1969). Although the focus of the exhibition was the Weißenhofsiedlung, a housing estate project for which the most progressive architects throughout Europe were invited to design buildings, "Die Wohnung" also acted as an important showcase for industrial design, with stands exhibiting the latest products from companies such as **AEG**. The modern tubular metal furniture designed by Mies van der Rohe, Mart Stam (1899–1986), Marcel Breuer (1902–1981), Le Corbusier (1887–1965) and others, which was used to furnish the interiors of the specially commissioned houses, was also widely publicized and revealed the increasing internationalism of the Modern Movement. Through this landmark exhibition, the Deutscher Werkbund succeeded in achieving a greater acceptance of Modernism both at home and abroad. The Werkbund was eventually disbanded in 1934, and though re-established in 1947, was by then a spent force. Through its activities and in particular its development of the concept of "types" – standardized designs that could be easily assembled from industrially produced components – the Deutscher Werkbund heralded the advent of large-scale industrial production and in so doing had an enormous impact on the evolution of German industrial design.

NIELS DIFFRIENT

BORN 1928
STAR, MISSISSIPPI, USA

Niels Diffrient studied aeronautical engineering at the Cass Technical High School, Detroit, and later trained at the Cranbrook Academy of Art, Bloomfield Hills, Michigan and at Wayne State University, Detroit. He worked in Marco Zanuso's Milan-based design studio from 1954 to 1955 and from 1946 to 1951 in the office of Eero Saarinen (1910–1961). In 1952 he joined **Henry Dreyfuss** Associates, becoming a partner in 1956. While there, he helped to compile data on anthropometrics which was eventually published in the three influential volumes entitled *Humanscale*. He also designed aircraft interiors as well as computers and X-ray equipment. In 1981 he founded Niels Diffrient Product Design in Ridgefield, Connecticut and has since designed ergonomically-conceived office systems furniture for Knoll and for Sunar-Hausmann. With his mastery of ergonomics, Diffrient is able to promote a highly resolved physical interaction between object and user. Regarding design as a "practical art", Diffrient attempts to provide superlative physical function in the most aesthetically pleasing way possible.

Diffrient chair, office seating system for Knoll International, 1979–1980

HENRY DREYFUSS

BORN 1904 NEW YORK, USA
DIED 1972 PASADENA, CALIFORNIA, USA

Henry Dreyfuss trained at the Ethical Culture School in New York prior to apprenticing with the industrial designer **Norman Bel Geddes**. While at Geddes' office from 1923 to 1929, he concentrated chiefly on theatrical work and designed costumes, sets and lighting for the Strand Theater, New York and for R. K. O.'s vaudeville theatres. Dreyfuss also worked as a consultant to Macy's before establishing his own design office in New York in 1929. From 1930 onwards he designed telephones for Bell Telephone Laboratories, including the *Model 300* (1937), the *Model 500* (1949) and the *Trimline* telephone (1964). Dreyfuss also created a model of "The City of Tomorrow" for **General Electric**, which was displayed at the 1939 New York World's Fair. Between 1938 and 1940, Dreyfuss designed two trains for the New York Central Railroad, including the streamlined *20th Century Limited* (1938). Later, in 1947, he executed a curious hybrid prototype, the *Convair* flying car for Vultee. Dreyfuss' straightforward business-like approach to the design process, which included working closely with engineers, contributed to the success of

Model 300 telephone for Bell Laboratories, 1937

his office. His large corporate clientele included AT&T, American Airlines, Polaroid, **Hoover** and RCA. His designs were characterized by the use of sweeping sculptural forms, and as such exemplified streamlining in American design. Like **Raymond Loewy**, Norman Bel Geddes and **Walter Dorwin Teague**, Dreyfuss re-styled many products for manufacturers so as to increase consumer demand through stylistic rather that technical innovation. Some of his designs bore a facsimile of his signature – an early example of designer labelling. Dreyfuss was a founder member of the Society of Industrial Design and the first president of the Industrial Designers Society of America. He was also a long-term faculty member of the engineering department at the California Institute of Technology. Dreyfuss' greatest contribution to industrial design practice, however, was his research into anthropometrics – the find-

←*Convair* car/plane for Consolidated Vultee Aircraft, 1947

←*Convair* car/plane for Consolidated Vultee Aircraft, 1947

ings of which were published in his influential books, *Designing for People* (1955) and *The Measure of Man* (1960). He also published a sourcebook of international symbols, which acknowledged the communicative power of symbols over words.

20th Century Limited locomotive for New York Central Railroad, 1938

Edison's original carbon filament lamp manufactured by the Corning Glass Works, 1879

THOMAS ALVA EDISON

BORN 1847 MILAN, OHIO, USA
DIED 1931 WEST ORANGE, NEW JERSEY, USA

Thomas Alva Edison pioneered some of the greatest and widest-ranging inventions of the 20th century, yet as a child his education was hampered by dyslexia and severe hearing problems. He was eventually removed from school and taught instead by his mother, who was a former schoolteacher. She gave him an elementary science book which outlined experiments that could be done at home. At the age of ten, Edison was allowed to set up his own science laboratory in the basement of his home. At the age of twelve, he began working as a trainboy for the Grand Trunk Railway and moved his laboratory into a baggage car so that he could do experiments during the layover periods in Detroit. The laboratory soon caught fire, however, and Edison lost his job. After this, Edison sold newspapers along the railroad and reputedly saved the life of a station official's son. The father was so grateful to Edison that he taught him how to use a telegraph. He later worked as a telegrapher in Toronto and, having to send a signal every hour, Edison devised a transmitter and receiver that could automatically telegraph a message even when he was asleep. He later moved to New York and began working on other inventions including a telegraph machine that could send multiple messages, an electric vote recorder and an improved stock market ticker, the rights to which he sold to the Gold & Stock Telegraphy Company for a massive $40,000 in 1870. By now wealthy, Edison established a workshop in Newark, New Jersey, to manufacture stock tickers, high-speed printing telegraphs and an improved version of Christopher Sholes' typewriter that was the first model suitable for commercial use.

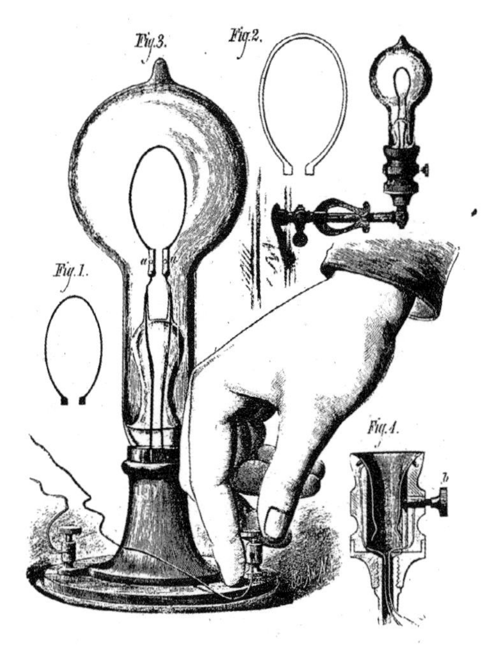

Drawing showing the elements of a light bulb, c. 1880

Thomas Alva Edison

In 1876 he built an "invention factory" in Menlo Park, New Jersey, and subsequently spent the majority of his time experimenting in this new science laboratory that had 60 employees. He worked at the laboratory for a decade and often had as many as 40 projects running at once. During this period, Edison was applying for up to 400 patents per annum. The numerous inventions conceived at Menlo Park and subsequently transformed into landmark products included the wireless telegraph (1875), the memograph (1876), an improved carbon-button telephone transmitter (1877), the phonograph (1877–1878), the incandescent electric light bulb (1878–1880) and the wireless induction telegraph (1885). Of these, the phonograph and the electric light bulb must be considered his greatest triumphs. The phonograph evolved from his research into telegraphy and was the first-ever recording machine. By developing a device that indented a strip of paper to record Morse Code, Edison observed that when the strip was run quickly, it emitted distinctive noises. The concept was adapted to sound recording by using a stylus that vibrated when exposed to sound, which in turn indented tin foil wrapped around a rotating drum. His incandescent electric light bulb was the result of many years of research and some 1,200 experiments. Using carbonized filaments made from burnt cotton thread, Edison's light bulbs could burn for up to 48 hours and were first used on the steamship *Columbia*. In 1878 he founded the Edison Electric Light Company to develop and eventually market this life-changing invention – the firm was incorporated as the Edison General Electric Company in 1892 and later became known as **General Electric**. In 1882 Edison established the first "electric light-power station", which enabled New York to become the first city to be illuminated by electricity. Another far-reaching discovery made by one of Edison's engineers in 1883 led to the development of the electron tube, which helped establish electronics as a completely new branch of science. Having founded an enormous new laboratory for the "business of inventing" in West Orange in 1887, which employed over 5,000 workers, Edison

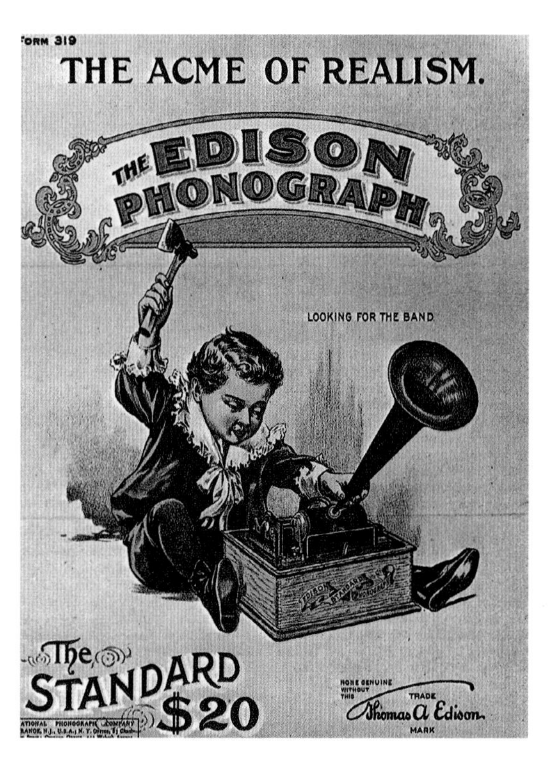

Poster advertising the Edison phonograph

Edison phonograph, c. 1880

went on to design improved versions of his phonograph, the first-ever motion picture camera (1891), the kinetograph and projecting kinetoscope for still and moving films (1897), the bipolar dynamo (1899) and the reversible galvanic battery (1900). Over his remarkable life, a total of 1093 patents were issued to Edison inventions. "All progress, all success, springs from thinking", he declared – but his phenomenal achievements were also the result of dogged perseverance and his ability to learn from his innumerable failures. As Edison put it so succinctly: "Genius is 1 % inspiration, and 99 % perspiration."

Edison bipolar dynamo generator manufactured by the Edison Machine Works, 1899

KENJI EKUAN

BORN 1929
TOKYO, JAPAN

Dubbed "the **Raymond Loewy** of Japan", Kenji Ekuan is Japan's most prominent industrial designer. He initially trained as a Buddhist priest before studying design at the National University of Fine Arts & Music, graduating in 1955. With fellow students he founded the GK (Groupe Koike) Design Group in 1953, which initially concentrated on industrial design. A scholarship from the Japan External Trade Organization (JETRO) enabled him to study industrial design at the Art Center College of Design in Pasadena, California, for one year. On his return to Tokyo in 1957 he became president of GK Industrial Associates. As well as designing motorcycles and audio equipment for Yamaha for over 40 years, GK has also created many other products, including packaging for Kikkoman, bicycles for Maruishi and cameras for Olympus. GK later expanded its remit to include graphic design, signage and urban planning. With his many publications on design practice, Ekuan has emphasized the need to humanize technology, democratize design and "seek the soul in material things". As both a world-class designer and a Buddhist monk, Ekuan has always pursued a harmonious connection between the material and spiritual worlds, and has been fascinated by the "communicative quality of design as a tangible medium".

Narita Express train for the East Japan Railways Company, 1991

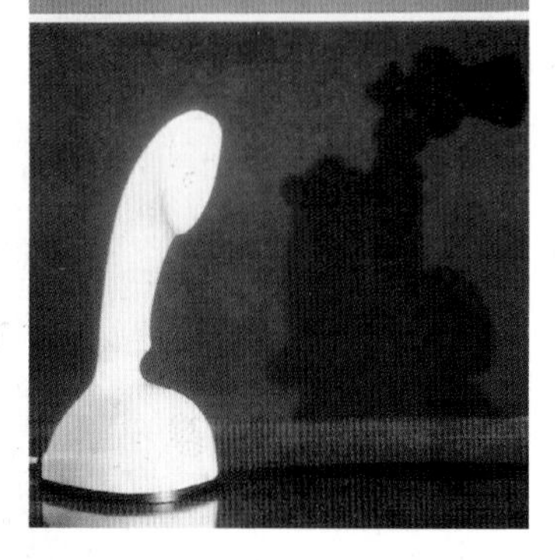

L. M. ERICSSON

FOUNDED 1876
STOCKHOLM, SWEDEN

Lars Magnus Ericsson (1846–1926) founded the L. M. Ericsson Company in Stockholm in 1876. The business initially concentrated on the repair of telegraph equipment, but by 1878 it was producing telephones that were based on an earlier design by **Alexander Graham Bell**. Soon after, L. M. Ericsson began manufacturing telephones of its own design, which exported throughout Europe, and which were the first to feature a combined transmitting/receiving handset. In 1909 Ericsson launched a cradle telephone, which was enormously influential throughout Europe. Seeking to update this landmark design, in 1930 the company commissioned the artist Jean Heiberg (1884–1976) and the Norwegian engineer Johan Christian Bjerknes to design a telephone with a Bakelite casing. Although not the first telephone to use plastics in its construction, this sculptural design was highly influential and inspired **Henry Dreyfuss**' later *Bell 300* model (1930–1933). Between 1940 and 1954, Hugo Blomberg (1897–1994), Ralph Lysell (1907–1987) and Gösta Thames (b. 1916) designed and developed another groundbreaking model for Ericsson, the *Ericofon* (launched in 1956). This unusual and forward-looking telephone had a completely unified form that integrated the earpiece, mouthpiece and dial. The design of the *Ericofon*, which incorporated new lightweight materials such as plastic, rubber and nylon, was made possible by the increasing miniaturization of technology. Fun yet stylish, it was the most popular one-piece telephone for over three decades. Today, L. M. Ericsson continues its commitment to innovative

Cover of the *Ericsson Review*, 1956

Model 88 telephone, 1909

Hugo Blomberg, Ralph Lysell & Gösta Thames, *Ericofon*, 1954 (launched in 1956)

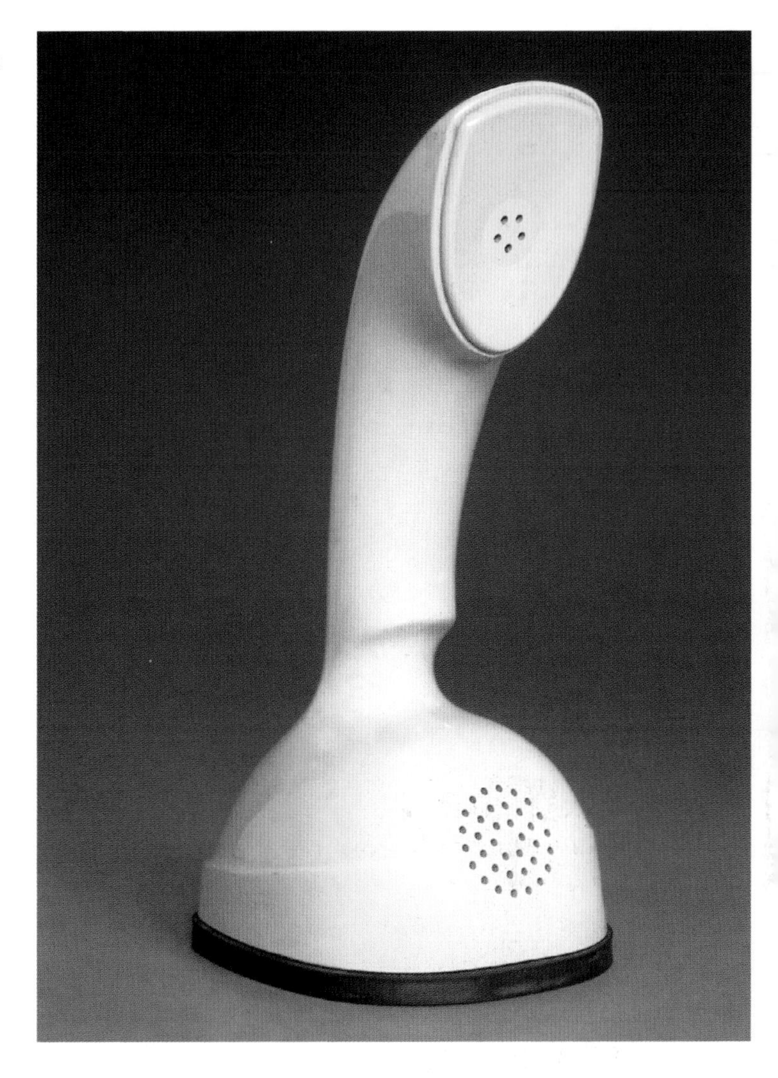

design and cutting-edge technology and remains one of the leading players in the telecommunications industry, with over 100,000 employees. The company is the world leader in mobile telephone systems, connecting nearly 40 % of the world's mobile callers. Ericsson is at the forefront of the WAP (Wireless Application Protocol) revolution – the bridge between mobile communication and the Internet – with landmark products such as the *R320*, the company's first mobile phone with a WAP browser.

HARTMUT ESSLINGER

BORN 1944
ALTENSTEIG, GERMANY

Hartmut Esslinger trained as an electrical engineer at the University of Stuttgart before studying industrial design at the Fachhochschule in Schwäbisch Gmünd. In 1969 he established his own industrial design consultancy, Esslinger Design, in Altensteig. His first client was the electronics company Wega Radio. Wega's subsequent acquisition by **Sony** in 1975 introduced Esslinger to the lucrative consumer electronic market that was booming in Japan. His sleek designs, such his *Concept 51K* hi-fi system for Wega (1975), brought him widespread recognition and were remarkable for their Modern dematerialist aesthetic. His work initially appeared to be influenced by the functionalist approach to design promoted by Max Bill (1908–1994) and **Hans Gugelot** at the Hochschule für Gestaltung in Ulm. It was, however, also driven by a rejection of "the prevailing Modern sentiments of 60s' Germany" and sought instead a "vision of what design could still yet become as op-

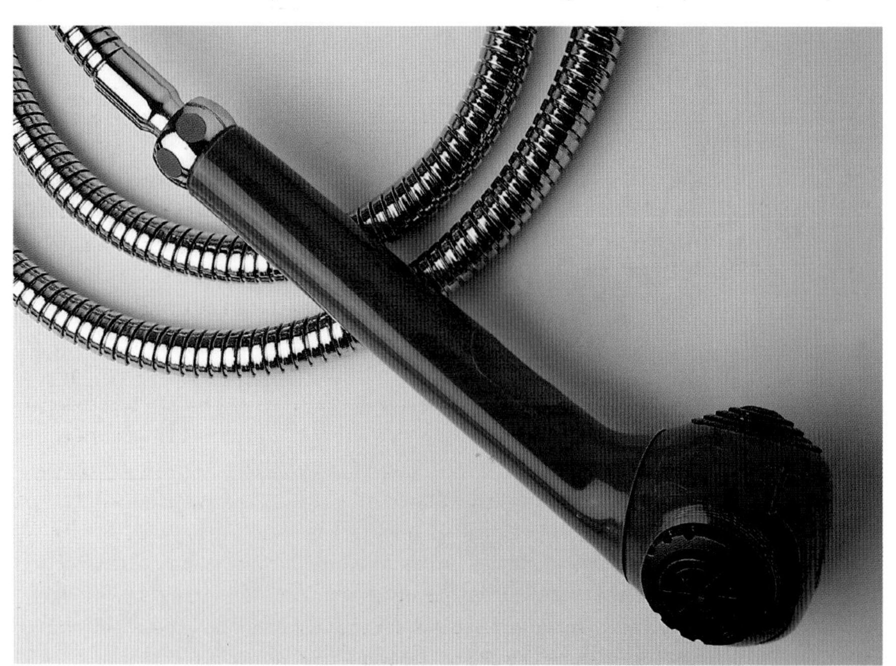

Esslinger Design, *Tri-Bel* showerhead for Hansgrohe, 1973

Esslinger Design, *Indusco Froller* rollerskates for Frollerskate, 1979Poster advertising the Fiat 500, c. 1936

posed to what it has already settled for being". In 1982 Esslinger renamed his consultancy Frogdesign (frog = Federal Republic of Germany) and opened an office in Campbell, California, so as to meet the design needs of the burgeoning computer industry in Silicon Valley. In 1984 the consultancy designed the off-white *Apple Macintosh* for **Apple Computer** and in so doing redefined the aesthetic parameters of personal computers. Two years later Esslinger opened a sister office in Tokyo, and during the 1990s established branches in technology hot-spots such as the "Silicon Prairie" of Austin, Texas, "Silicon Alley" in New York, "Media Gulch" in San Francisco and "Silicon River" in Düsseldorf. Frogdesign has also designed cameras, synthesizers, binoculars, communications equipment and office seating for such companies as RCA, **Kodak**, Polaroid, Motorola, Seiko, **Sony**, Olympus, AT&T, **AEG**, König und Neurath, Erco, Villeroy & Boch, Rosenthal and Yamaha. Esslinger attempts to humanize technology with sculptural forms or visual references to create more user-friendly products. He has always had an international outlook and recognizes the "benefit of bringing Asian teamwork, European discipline, German precision, American optimism and Californian craziness together" to produce a very unique and culturally melding creative enterprise.

Dante Giacosa,
Fiat 500, 1936

FIAT

FOUNDED 1899
TURIN, ITALY

Fiat is Italy's largest automotive company and one of the world's largest industrial groups, with diversified production ranging from buses and trains to farm tractors and aircraft. Founded in 1899 as the Fabbrica Italiana Automobili Torino by a group of investors including Giovanni Agnelli (1866–1945), Fiat received instant recognition for its luxury cars as well as its racing triumphs. The company grew rapidly in terms of both sales and production, and built its first American factory in 1909 in Poughkeepsie, New York. Later, in an attempt to increase production, Agnelli decided to build the largest automotive plant in Europe, which opened in Lingotto in 1922. Implementing assembly-line methods of production, Fiat used this factory to transform the car from a relatively exclusive item into a much more accessible product, a philosophical approach to manufacturing that the company has adhered to throughout its history. Its most successful pre-war car was the compact Fiat 500, or "Topolino" (Little Mouse) as it became known, which was designed by Dante Giacosa (b. 1905). Launched in 1936, the *500* was the smallest mass-produced car in the world and over 519,000 models were produced between 1936 and 1955. Its rounded lines broke with the tradition of box-shaped forms, while its light metal body enabled extremely fluid pro-

→Fiat *Nuova 500A*, 1957

duction. During the post-war period, Fiat's objective was to produce the sort of cars that the large American manufacturers were not making – cars with smaller engines that everyone could afford. This led directly to models such as the Fiat *600* (1955) and the Fiat *Nuova 500* (1957). These inexpensive and diminutive vehicles, together with their variations, were manufactured in their millions and did much to democratize car ownership in Italy. They also became potent symbols of Italy's economic miracle. In 1978 Fiat introduced the world's first flexible robotic assembly lines in its plants in Rivalta and Cassino, and a year later merged with Lancia, Ferrari, Autobianchi and Abarth to form Fiat Auto SpA. During the 1980s and 1990s, the company continued producing "people's cars" such as the Fiat *Panda* (1980), the Fiat *Uno* (1983), which was styled by Giorgetto Giugiaro and became Fiat's best-ever selling car, and the intermediate sized Fiat *Punto* (1993). In 1999 Fiat launched the versatile 6-seater *Multipla*, a groundbreaking mid-range MPV with non-conformist styling that was voted Car of the Year 2000 by the BBC. Highly innovative designs such as these are developed in Fiat's Advanced Design centre, which employs a youthful team devoted specifically to generating new car concepts.

Fiat *Multipla*, 1999

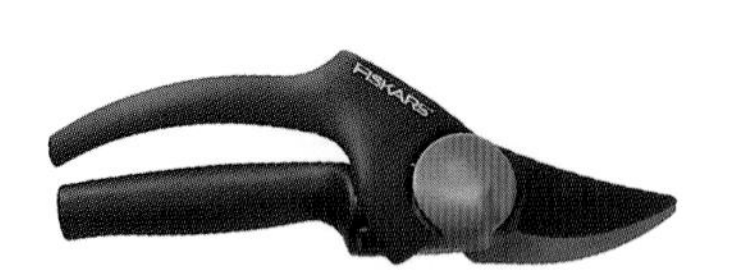

FOUNDED 1649
FISKARS, FINLAND

Olavi Lindén, Clippers secateurs, 1996

As the oldest industrial company in Finland, Fiskars can trace its origins to an ironworks established by Peter Thorwöste in southern Finland in 1649. From the 1820s, Fiskars was forging high-quality tableware, kitchen knives and scissors, but it was not until 1967 that the company came to the world's attention when it launched its famous orange-handled *O-series* scissors (1963). Developed by the engineer and wood-carver Olof Bäckström (b. 1922), the prototype for the ergonomically conceived ABS handles of these scissors was carved from wood. Licensed worldwide, the scissors were also manufactured as pinking shears and in a left-handed version. Over the succeeding years this benchmark ergonomic design, now known as the *Classic*, has been developed into a family of 18 scissors. More recently, under the guidance of Olavi Lindén (b. 1946), the Fiskars design team has developed a range of garden tools that has been awarded numerous design prizes. Known as *Clippers* (1996), these tools reflect Fiskars' "commitment to quality and systematic application of ergonomic principle, which result in products that are easy, safe and pleasurable to use".

Olavi Lindén, *Handy* axe and *Clippers* secateurs, 1994 & 1996

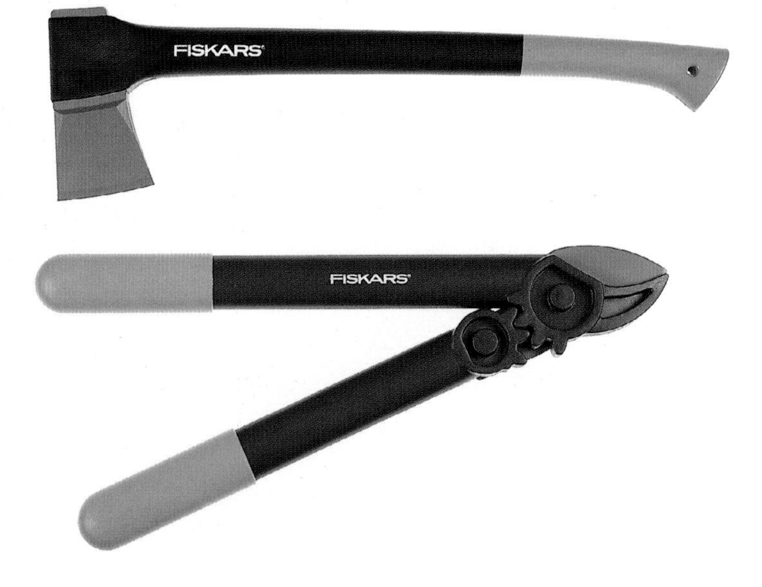

FORD

FOUNDED 1903
DETROIT, MICHIGAN, USA

Henry Ford I

While Henry Ford (1863–1947) did not invent the automobile, it was his vision that made the car accessible to literally millions of people. His primary goal was to "build a motor car for the great multitude ... it will be so low in price that no man ... will be unable to own one." In the 1860s, two events occurred that enabled his dream to be fulfilled – the invention of the open-hearth process in 1864, which heralded the birth of modern steel manufacture, and the oil industry's laying of a pipeline in the Allegheny River valley in 1865, which signalled the beginning of an enormous network of petrol stations that would eventually fuel Ford cars. Funded by twelve investors, the Ford Motor Company was established in June 1903 in a converted Detroit wagon factory. The first car was sold only a month later and was described as "the most perfect machine on the market.... so simple that a boy of fifteen can run it". Over the following five years, Henry Ford directed a research and development programme that resulted in a plethora of models.

Model T, 1925

↑*Model T*, 1912

↗*Model T*, 1913

↑*Model T*, 1923

↗*Model T*, 1927

These vehicles, which were coded alphabetically, included two-, four- and six-cylinder cars, some of which were chain driven while others were shaft driven. Not every model made it into production, and of those that did, not all were successful: while the *Model N*, which retailed for a modest $500, sold well, for example, the expensive *Model K* limousine did poorly. Following the failure of the *Model K*, Henry Ford insisted that the company's destiny lay in the manufacture of inexpensive cars. At this time, George Selden held a patent for "road locomotives" powered by internal combustion engines, and most car manufacturers paid him royalties. Ford, however, believed Seldon's claim of exclusivity to automobile technology was invalid and refused to pay any royalties. In 1911, after many years of costly litigation, Ford finally won the suit brought against him by Selden and in so doing freed the entire automobile industry from what was, clearly, an impediment to its progress. Despite this legal distraction, the Ford Motor Company continued to expand and in October 1908 the pivotal *Model T* came into being. Described by Henry Ford as the "universal car", the low-cost and reliable *Model T* became an instant success. The *Model T* was so much in demand that Henry Ford instigated a system of compartmentalized production and came up with the concept of a moving assembly line. This approach to manufacturing, which became known as Fordism, dramatically reduced assembly time and resulted in over 15 million *Model T*s being built between 1908 and 1927. The succeeding *Model A* (named after Ford's first car) placed

Ford *Escort GT*,
1967

Ford *Fiesta 1.6*,
1976

Ford *Focus*,
1999

greater emphasis on safety and comfort than its predecessor and became known as "the baby Lincoln" because of its softer contours. Although sales were impacted by the Great Depression, some 5 million *Model A*s were produced before it was succeeded in 1932 by the *V-8*. In 1948 Ford launched the *F-Series*, a pick-up truck that was "built stronger to last longer". Becoming the best-selling vehicle in North America (a title it still holds) – over 27 million *F-Series* trucks have been manufactured to date. In Europe, Ford became better known for its practical "everyman" cars, such as the *Escort* and *Fiesta*. During the late 1990s, the company evolved a new language of design known as "New Edge Design", which combines "smooth, sculpted surfaces with clear, crisp intersections". This fresh approach led to such innovative vehicles as the *Ka*, *Puma*, *Cougar* and *Focus*, which was voted Car of the Year in Europe in 1999 and in America in 2000. Ford's commitment to design innovation resulted in it commissioning the product designer Marc Newson (b. 1963) to develop a concept car for the 21st century, the Ford *012C*, which features a number of imaginative details such as a slide-out luggage tray and a single LCD headlight. Ford also acknowledges its heritage and its new retro design *Thunderbird* is an intriguing reinterpretation of the legendary 1950s original. Like all motor manufacturers, Ford has become increasingly mindful of environmental issues and is committed to developing a new generation of vehicles that radically reduce CO_2 emissions.

Ford *Ka*, 1996

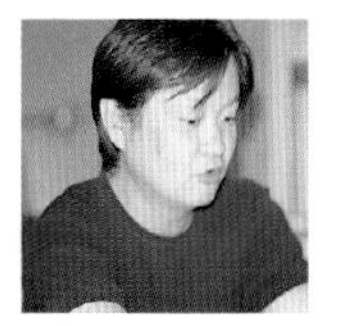

NAOTO FUKASAWA

BORN 1956
KOFU, JAPAN

Naoto Fukasawa trained as a product designer at Tama Art University and later worked as the chief designer of the R&D Design Group at the Seiko Epson Corporation in Japan. Since 1989 Fukasawa has worked for ID Two (later **IDEO**), leading a design team that has formulated a coherent visual language for NEC products ranging from computer notebooks to LCD projectors, and which have received several G Mark and Hanover iF awards. He has also assisted **Apple Computer** in developing a new vocabulary of design in which the ergonomic form of the product conforms to the user's actions. Fukasawa's *Left Ventricular Heart Assist* (1990), a computer that externally controls an artificial heart for a recovering patient, is not as he describes "a clean, health-goods type of design" but a no-nonsense and robust design that ensures the continuous beating of the heart. Fukasawa, who was design director of IDEO Japan, is fascinated by the interaction between man and technology. He has collaborated with Sam Hecht (b. 1969) on a number of products that attempt to humanize computer technology, and which are born out of his belief that "perfection is admitting to oneself the existence of imperfection."

Left Ventricular Heart Assist for Baxter, 1990 – co-designed with Sam Hecht

RICHARD BUCKMINSTER FULLER

BORN 1895 MILTON, MASSACHUSETTS, USA
DIED 1983 LOS ANGELES, CALIFORNIA, USA

Richard Buckminster Fuller studied mathematics at Harvard University from 1913 to 1915 and in 1917 enrolled at the US Naval Academy in Annapolis. While at the Academy he began his "theoretical conceptioning", which included a proposal for "flying jet-stilts porpoise" transport that was eventually published in 1932. After leaving the military, he founded the Stockade Building Company in 1922, which was a financial failure and led to his personal bankruptcy. He committed himself to finding global solutions to social problems after the death of his four-year-old daughter in 1922, which he believed was due to inadequate housing. In 1933/34 he founded the 4D Company in New York to develop his design concepts, which were driven by his ambition to evolve a "design science" that would bring about the best solutions with the minimum use of energy and materials. He based this concept on the Modern Movement principle of getting the most with the least and named it "Dymaxion", which was derived from "dynamic" and "maximum efficiency". In 1929 he launched the magazine *Shelter* and was its publisher and editor from 1930 to 1932. From 1932 to 1938 he was director and chief engineer of the Dymaxion Corporation, which he set up to develop and produce three streamlined prototype cars that were based on his dymaxion principles and were inspired by aircraft design. It was claimed by Fuller that the Dymaxion car of 1934 could accelerate from 0 to 60 mph in three seconds and provide 30 mpg fuel consumption. The prototype car, however, was not progressed due to several serious design flaws. From 1927, Fuller also developed the Dymaxion House concept and between 1944

Dymaxion car, 1932

Dymaxion Dwelling Machine (Wichita House), 1944–1947

and 1947 a prefabricated metal dwelling known as the Wichita House. Although the company he set up to manufacture this industrially produced architecture received an astonishing 38,000 orders after its press launch, he was not prepared to commence manufacture of the Wichita House until its design had been perfected. His backers subsequently grew disheartened with the time it was taking and this product architecture project was shelved. Fuller's most famous invention was the geodesic dome of 1949, which had a wide range of applications from industrial to military to exhibitions. The geodesic dome remains the only large dome that can be set directly on the ground as a complete structure. It is also the most economic space-filling structure and the only practical type of building that has no limiting dimensions. The outcome of Fuller's "more for less" approach, the geodesic dome employed a minimum of materials and could be easily transported and assembled. This remarkable design offered a means of producing ecologically efficient housing for the mass-market, and Geodesics Inc. was set up in 1949 to develop the concept. Fuller was also the first person to coin the expression "spaceship Earth". Prolific communicator, humanist, polymath and futurist, Buckminster Fuller believed that the creative abilities of humankind were unlimited, and that technology and comprehensive, anticipatory design-led solutions could eliminate all barriers to humanity's expansion into a positive future.

NORMAN BEL GEDDES

BORN 1893 ADRIAN, MICHIGAN, USA
DIED 1958 NEW YORK, USA

Norman Bel Geddes studied art at the Cleveland Institute of Art and later trained briefly at the Art Institute of Chicago. In 1913 he worked as a draftsman in the advertising industry in Detroit, where he designed posters for Packard and **General Motors**. In 1916 he wrote a stage play and subsequently worked as a theatre-designer for six productions in Los Angeles. In 1918 he became a set designer for the Metropolitan Opera Company in New York, before moving to Hollywood in 1925, where he designed lavish film sets. Influenced by Frank Lloyd Wright (1867–1959) and the German Expressionist architect Erich Mendelsohn (1887–1953), in 1927 Geddes turned to architecture and product styling. While he was a highly successful industrial design consultant – notably working for the Toledo Scale Company in 1929 and the Standard Gas Equipment Corporation in 1931 – he is best remembered as a design propagandist. In 1932 he published a book entitled *Horizons*, in which he outlined his approach to industrial design and his belief in the supremacy of the teardrop shape. Geddes also designed futuristic cars for the Graham-Paige automobile company (1928) and streamlined products such as radios for Philco (1931), radio casings for RCA, metal bedroom furnishings for Simmons (1929) and the well-known *Soda King Syphon* for the Walter Kidde Sales Co. (1932). One of his most notable achievements was his standardization of kitchen equipment. The design of his modular and streamlined *Oriole* stove (1931) for the Standard Gas Equipment Company was inspired by the construction of skyscrapers and, as such, had a steel frame onto which white vitreous enamel panels were clipped. Geddes also designed the General Motors' "Futurama" display for the 1939 New York World's Fair, which speculated on the future (as it was believed it would be in 1960) and correctly predicted the freeway system. Geddes was one of the greatest exponents of streamlining and was also an important pioneer of industrial design consulting.

Soda King Syphon for the Walter Kidde Sales Company, 1932

GENERAL ELECTRIC

FOUNDED 1892
ALBANY, NEW YORK, USA

Complete line of heating devices, 1907

The origins of this phenomenally successful company can be traced to the inventor **Thomas Alva Edison**, who established the Edison Electric Light Company in 1878 and a year later invented the first practical electric light bulb. This venture, renamed the Edison General Electric Company in 1889, eventually merged with the Thomson-Houston Electric Company in 1892 to form the General Electric Company. Following Edison's earlier idea of establishing "invention factories", the company founded the GE General Engineering Laboratory in 1895 to conduct research into advanced engineering and instrumentation. In 1900 the company established the GE Research Laboratory, one of the first industrial laboratories in America to undertake basic scientific research. These laboratories were crucial to the development of many inventions and discoveries, including the high-frequency alternator (1906) devised by Dr. Ernest Alexanderson (1878–1975) which revolutionized radio broadcasting, the gas-filled incandescent lamp (1913) invented by Dr. Irving Langmuir (1881–1957) that was the forerunner of the light bulbs still used today, and the first practical X-ray tube (1913) invented by Dr. William Coolidge (1873–1975), which was the predecessor of the modern medical X-ray tube. During the late 1920s General Electric also pioneered television broadcasting in America, using technology developed by Alexanderson. In addition, the company was renowned for its labour-saving domestic appliances, from electric heaters and washing machines to vacuum cleaners and refrigerators – all of which literally transformed the

Model AW1 washing machine, 1933

The first home television reception at the Schenectady residence of Ernest Alexanderson, c. 1928

Arthur BecVar, redesign of cylinder vacuum cleaner for General Electric, c. 1950

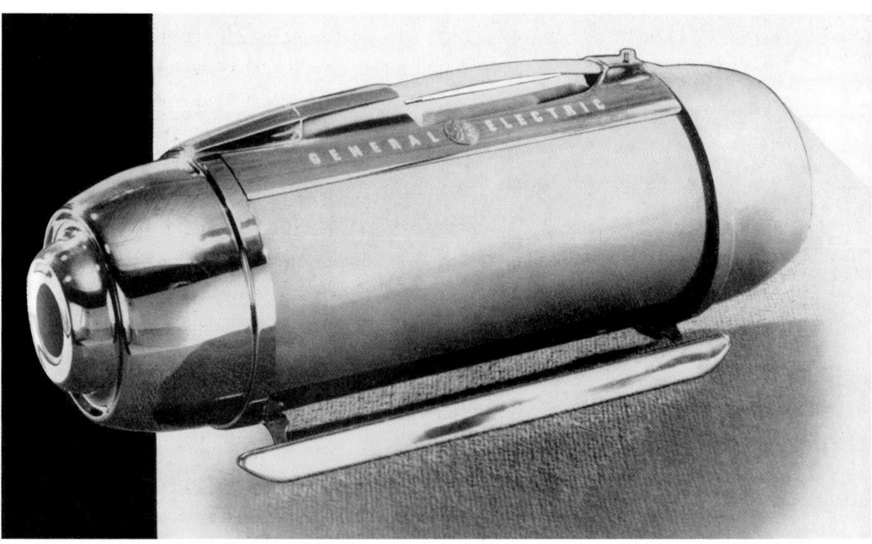

GE satellite station in California, c. 1992

American way of life. In the 1940s, GE produced engines that powered the first American jet-propelled aeroplane, the Bell *P-59 Airacomet* (1942) and the world's then fastest aeroplane, the Lockheed *P-80 Shooting Star* (1945). The company later produced the *J-79*, the first jet engine to propel an aircraft at twice the speed of sound. During the 1960s, GE contributed its technical expertise to NASA's Apollo moon-landing mission (1969) and developed weather satellites. In 1975 its research labs designed an essential component for the development of the CAT (computerized axial tomography) scanner and during the 1980s acquired the medical diagnostic imaging company, CGR. Continuing its commitment to television broadcasting, GE purchased RCA in 1986, which included the NBC television network, and subsequently launched two new networks, CNBC (1989) and MSNBC (1996). For over a century, General Electric has been driven by technological innovation and its life-changing and life-enhancing inventions have touched us all. Central to its continuing success are the corporation's twelve multi-disciplinary laboratories that reflect the diversity of its operations: from lighting systems to satellite systems, from aircraft engines to washing machines, from state-of-the-art plastics to medical diagnostic equipment. The research and development currently being undertaken in General Electric's laboratories will almost certainly shape the future of industrial design, just as it has done over the past 100 years.

Oldsmobile, 1903

GENERAL MOTORS

FOUNDED 1908
HUDSON, NEW JERSEY, USA

General Motors is the world's largest automotive manufacturer and throughout most of the 20th century was the world's largest industrial corporation. For this reason, it has historically been regarded as a barometer of the American economy. Seven years before its incorporation in 1908, the company manufactured the first car to be produced in quantity in America. The year of its official founding, GM launched the first electric headlamp, and in 1910 introduced "closed bodies' as standard equipment. In 1912 the company pioneered the first all-steel car body, which provided much greater strength and safety than earlier automobiles. In 1924 GM established the first automotive proving-ground test facility in Milford, Michigan. During the early 1920s, the company's chairman predicted the increasing importance of styling in the automotive industry and subsequently commissioned Harley Earl (1893–1969) to work on the 1927 La Salle, the first mass-produced car to be developed by an automobile stylist. In 1928 GM established its own

La Salle, 1934

Chevrolet, 1955

styling department, which was headed by Earl and was the first of its kind. This influential in-house facility devised annual stylistic changes in an effort to accelerate the aesthetic life cycle of automobile models and, as a result, increase sales. As well as instigating this programme of annual cosmetic changes, GM continued to produce real innovations, such as the first built-in trunk (1933), which revolutionized the overall form of the car. Under Earl's guidance, GM also developed its first concept car, the Buick *Y-Job* (1938). By 1940 GM had produced a massive 25,000,000 vehicles, which sheds some light on the extent of pre-war car ownership in America. After the war GM introduced several innovations to its product line, including the "airfoil' fender (1942–1948), the curved windscreen (1948) and the modern, high-compression overhead valve V-8 engine, the first of which were used on the company's Cadillac and Oldsmobile models. The famous Chevrolet *Corvette* launched in 1953 became the first GM concept car to have evolved unaltered from a Motorama auto show to a full-scale production model. Although GM had set up the first safety-testing laboratory in 1955, the fateful Chevrolet *Corvair* which it produced in 1960 was famously attacked for its lack of safety and tendency to roll over by the young lawyer, Ralph Nader, in his book *Unsafe at any Speed* (1965). Nader's legal action against GM and his subsequent victory led the US Congress to pass 25 pieces of consumer

Corvette, 1963

legislation between 1966 and 1973, which opened the floodgates for product-liability law suits in America. Learning from this difficult experience, GM introduced the first energy-absorbing steering column (1966), the first side-guard door beam (1969) and the first airbags in production vehicles (1974). In 1975 it also began fitting the first-ever catalytic converters on all its cars sold in America. In more recent times, in 1996 GM launched the first modern from-the-ground-up electric car, the *EV1*. This remarkable zero-emission vehicle, with its sophisticated lightweight aluminium body, has the most aerodynamic shape of any production car ever. It is powered by a 137 hp, 3-phase AC induction motor and is offered with two battery technologies: an advanced high-capacity lead acid and an optional Nickel Metal Hydride, which doubles the *EV1*'s range from 55–95 miles to 75–130 miles. Today General Motors boasts many major automobile brands, including Chevrolet, Chevy Trucks, Pontiac, Oldsmobile, Buick, Cadillac, GMC and Saturn, and employs over 397,000 people with operations in 73 countries. As a truly global corporate giant and with the will to innovate such revolutionary vehicles as the *EV1*, General Motors seems likely to continue to dominate the automotive industry as it has done for nearly 100 years.

EV1, 1996

GILLETTE

FOUNDED 1901
BOSTON, MASSACHUSETTS, USA

King Camp Gillette

Reared in Chicago, King Camp Gillette (1855–1932) became the star salesman at the Crown Cork & Seal Company, which mainly manufactured cork bottle caps. His mentor at the company, who noticed his predilection for mechanical tinkering, suggested that he "try to invent something like the Crown Cork product which, when used once, is thrown away" so that the customer would keep coming back. Gillette became obsessed with this idea and one morning in 1895, while struggling to hone his straight razor, he had a vision of a razor with a separate handle that clamped a thin double-edged disposable steel blade between two plates. After six difficult years, in 1901 Gillette eventually founded the American Safety Razor Company, which in 1902 became the Gillette Safety Razor Company. While not the first "safety razor", Gillette's innovative refillable razor and blade system offered much greater convenience. It was first manufactured in 1903 and patented a year later. During the First World War, the company supplied 3.5 million razors and 36 million blades to the US armed forces. In 1932 the company introduced the famous double-edged *Blue Blade*, which dominated the blade market for many years. Gillette diversified its interests with its acquisition of the Paper Mate Pen Company in 1955 and **Braun** AG in 1967. As a leading

↘Early Gillette safety razor, c. 1904

Patent drawing of King Camp Gillette's first safety razor, 1901 – the final production model varied slightly

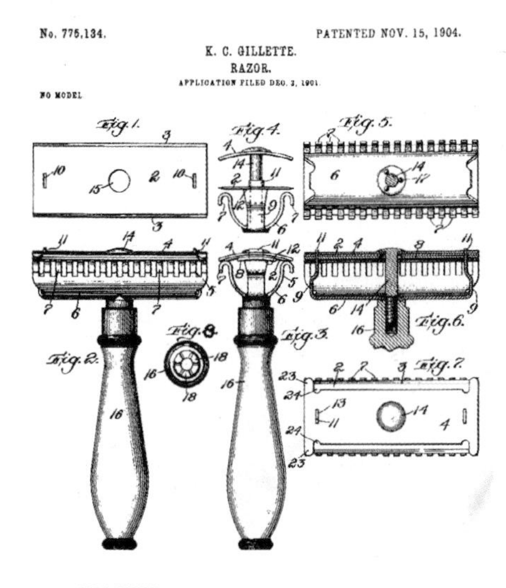

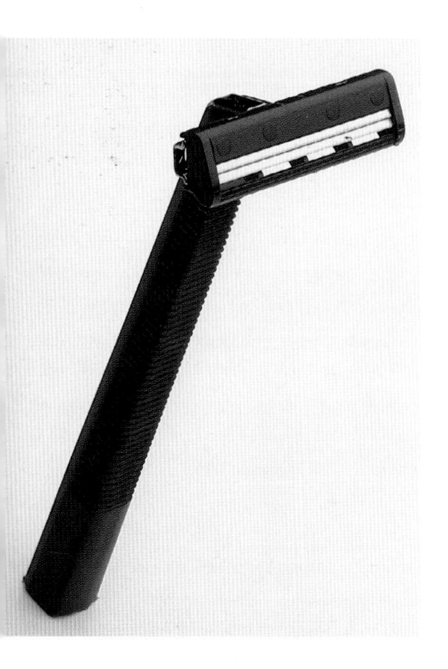

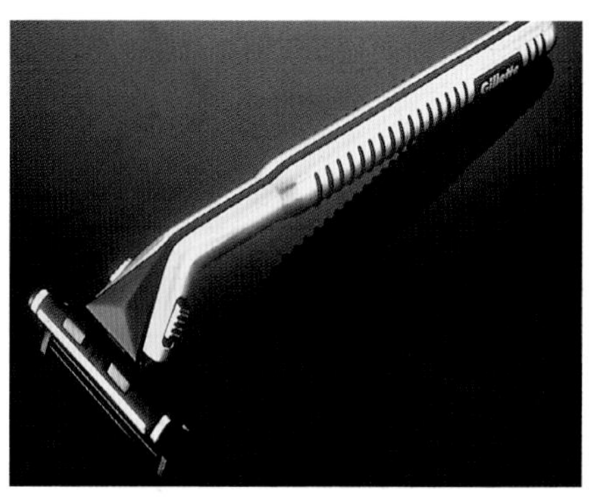

Trac II, 1971

↗*Sensor*, 1990

→*Mach 3*, 1998

manufacturer of electric razors, Braun was renowned for its product design team, which was headed by Dieter Rams (b. 1932), and which now began developing products for Gillette. In 1971 the company introduced the world's first twin-bladed cartridge system, the *Trac II*, which became an immediate best-seller. Later, after ten years' development, Gillette launched the *Sensor* system (designed by the Braun design team) in 1990. This revolutionary product featured thin spring-mounted blades set in a pivoting plastic cartridge. In the year of its introduction, 24 million razors and an astonishing 350 million cartridges were sold. As the most successful product launch in Gillette's history, the lessons learned from *Sensor* were clear: it reaffirmed the company's commitment to "spend whatever it takes to gain technological supremacy in a category, and then produce innovative products that will capture consumers, even at a premium price". This philosophy led directly to Gillette's latest flagship product, the remarkable triple-bladed *Mach 3*. By far the most expensive wet razor systems product today, the ultra-thin *Mach 3* blades were the first ever to be coated with a diamond-like film. Throughout the 1980s and 1990s, Gillette continued diversifying by purchasing Oral-B, Waterman, Parker Pen and Duracell, with the result that, by the year 2000, it had become the world's second most profitable brand after Coca-Cola.

KENNETH GRANGE

BORN 1929
LONDON, ENGLAND

Kenneth Grange trained at the Willesden School of Arts & Crafts in London between 1944 and 1947. Later, while completing his national service, he was trained as a technical illustrator by the Royal Engineers. He subsequently worked as an assistant for various architecture and design practices before founding his own London-based design office in 1958. Two years later, he redesigned a food mixer for Kenwood that had been originally launched in 1950. His resulting *Chef* food mixer (1960), which was restyled in just four days, was inspired by earlier kitchen appliances designed by Gerd Alfred Müller (b. 1932) for **Braun**. Grange worked as a design consultant to Kenwood for over 40 years and regularly updated this classic British design. In 1972 Grange, together with Theo Crosby, Alan Fletcher, Colin Forbes and Mervyn Kurlansky formed the design partnership Pentagram. Among Grange's many notable designs as part of Pentagram are the **Kodak** *Pocket Instamatic* camera (1975), the Parker 25 range of pens (1979), the *Royale* and *Protector* wet razors for Wilkinson Sword (1979 & 1992) and the *ST50* travel steam iron for Kenwood (1995). Grange also designed the exterior body of British Rail's *125 Intercity* high-speed train (1971–1973), Adshel bus shelters for London Transport (1990) and the restyled classic London black cab for London Taxis International (1997). Grange has received numerous awards for his esign work, including ten Design Council awards and the Duke of Edinburgh's Prize for Elegant Design in 1963. From 1985 to 1987 he served as

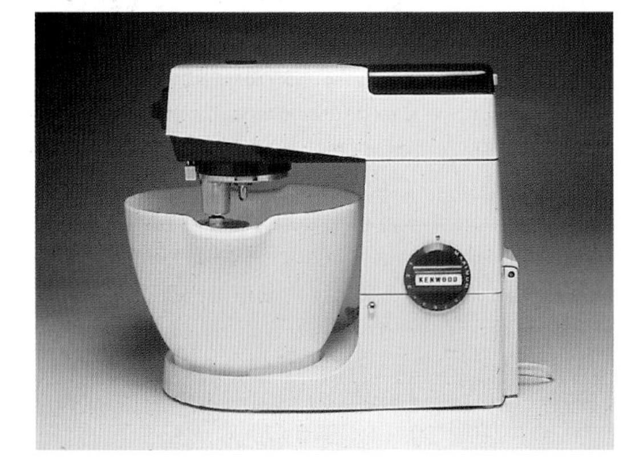

Kenwood *Chef A701* food mixer, 1960

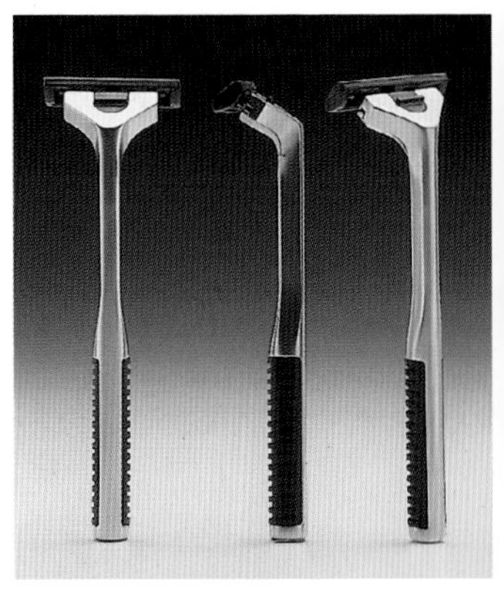

↘*Royale* razors for Wilkinson Sword, 1979

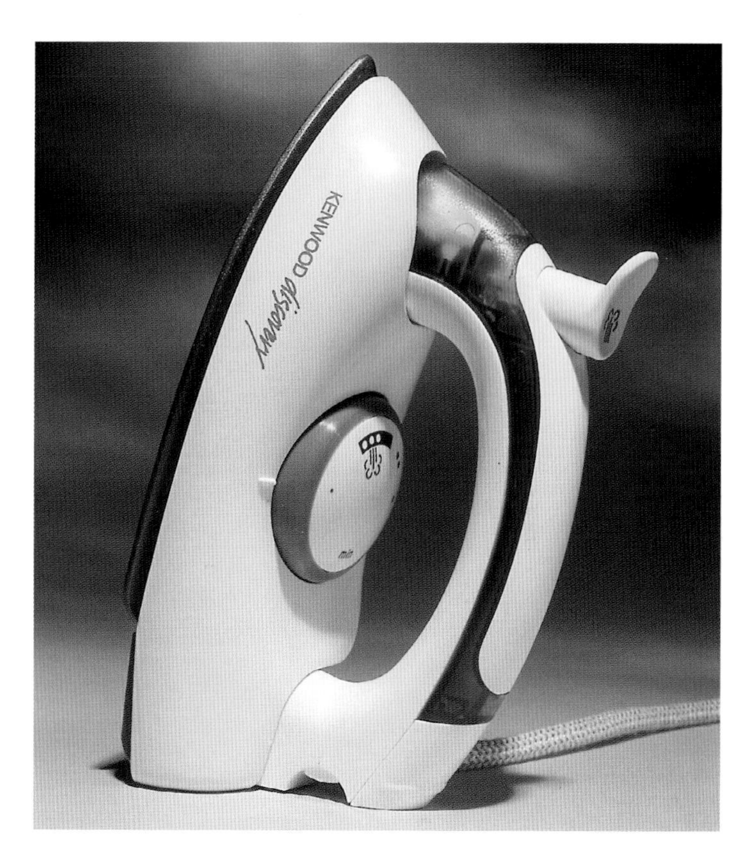

ST50 travel steam iron for Kenwood, 1995

Master of the Faculty of Royal Designers for Industry and was appointed president of the Chartered Society of Designers in 1987. His work was also the subject of a one-man show at the Boilerhouse at the Victoria and Albert Museum, London, in 1983. In 1997 he retired from Pentagram and established his own independent design office. Grange regards design as a means of innovation and believes that design should be an integral part of the manufacturing process. Grange combines the traditional attributes of British design – honesty, integrity and appropriateness – to create high-quality, meticulously-detailed products that meet the criteria for "good design". Throughout his long career Grange has shown a remarkable consistency – his gently elegant designs have all had an underlying robustness – which reflects his skill in balancing functional requirements with those of aesthetics. In many ways, Grange's work eloquently defined British design during the last half of the 20th century.

HANS GUGELOT

BORN 1920 CELEBES, INDONESIA
DIED 1965 ULM, GERMANY

After studying architecture, Hans Gugelot worked with Max Bill until 1954. From 1954 to 1965 he worked in the design department at **Braun**, where he helped develop a house style with a strong visual identity based on functionalism and essentialism. Between 1954 and 1965 Gugelot also directed the product design department 2 at the Hochschule für Gestaltung in Ulm, where he promoted the "form follows function" approach to design. Among his best-known products, the *Phonosuper SK 4* radio-phonograph (1956) was co-designed with Dieter Rams (b. 1932) and nicknamed "Snow White's Coffin" because of its clear acrylic lid and hard-edged geometric formalism. Other notable designs for Braun included the *Sixtant* electric shaver (1962). Gugelot also designed the *S-AV 1000* carousel slide projector for **Kodak** (1962) and built-in furniture for Bofinger (1954). While Gugelot's career was short, his influence – particularly upon the development of German product design – was immense.

Sixtant electric shaver for Braun, 1962 (co-designed with Gerd Alfred Müller)

D. J. De Pree

HERMAN MILLER

FOUNDED 1923
ZEELAND, MICHIGAN, USA

Robert Propst (b. 1921) wrote of Herman Miller: "Our aspiration has always been to design things that are fashion-proof; that have more quiet, enduring qualities; that try to retreat to more elementary qualities". The origins of this remarkable company can be traced back to 1923, when D. J. De Pree and his father-in-law, Herman Miller, acquired the majority holding in the Michigan Star Furniture Company and subsequently renamed it the Herman Miller Furniture Company. Like many furniture manufacturers at the time, Herman Miller produced reproduction furniture that was popular within the mainstream market. In 1930, however, when De Pree met the designer Gilbert Rohde (1894–1944), the company's destiny changed. Rohde proposed a high quality range of Modern furniture that was of the utmost simplicity, so that all the value went into materials and construction rather than decorative surface treatment. The subsequent success of Rohde's Modern furniture led Herman Miller to abandon the manufacture of reproduction furniture altogether in 1936, and by 1941 the company had opened a New York showroom to display its new Modern designs. In 1946 George Nelson (1907–1986) succeeded Rohde as design director and subsequently brought in other talented designers, including Charles Eames (1907–1978), Isamu Noguchi (1904–1988)

Robert Propst, *Action Office II* system, 1968

Don Chadwick & Bill Stumpf, *Aeron* office chair, 1992

and Alexander Girard (1907–1993), to create high-quality Modern furnishings for the company. Over the following years, Herman Miller produced designs that met De Pree's criteria for good design – "Durability, Unity, Integrity, Inevitability". These included Charles and Ray Eames' moulded plywood chairs (1945–1946) and plastic shell series of chairs (1948–1950) and George Nelson's *Comprehensive Storage System* (1959) and *Action Office I* (1964–1965). 1968 saw the launch of Robert Propst's landmark *Action Office II*, which literally transformed the office landscape. Since then, the company has been a leader in the contract furniture market and has produced a series of pioneering office chairs by Bill Stumpf (b. 1936) and Don Chadwick (b. 1936) – the *Ergon* (1976), the *Equa* (1984) and the revolutionary *Aeron* (1992). As one of America's most admired corporations, Herman Miller continues to flourish through its quest for design and manufacturing excellence.

HOOVER

FOUNDED 1908
NORTH CANTON, OHIO, USA

Vacuum cleaner for the Electric Suction Sweeper Company, 1908

The first Hoover vacuum cleaner was based on a design by a department store janitor, James Murray Spangler (1848–1915). He realised that if the sweeping action of a carpet cleaner was combined with the relatively new idea of using suction to remove dirt and dust from carpets, performance would be dramatically improved. His wood and tin prototype used a pillowcase for the bag and although it was a relatively clumsy device, it worked fairly well. Lacking the funds to commercialize this new invention, Spangler managed to interest his friend, W. H. Hoover (1849–1932), who was a manufacturer of leather goods. In 1908 the Electric Suction Sweeper Company (later re-named the Hoover Suction Sweeper Company in 1910 and then The Hoover Company in 1922) was established and Spangler became its production manager. Frank Mills Case subsequently re-designed the cleaner, giving it an aluminium casing. These first electric vacuum cleaners were initially produced at the rate of six to eight machines a day and weighed nearly 40 pounds, but the development in 1909 of a small high-speed universal motor by Hamilton & Beach,

Advertisement for the Hoover vacuum cleaner *Model 150*, 1936

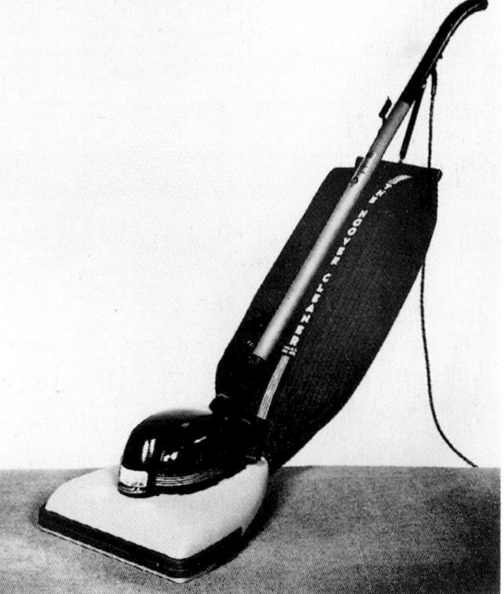

↘Henry Dreyfuss, Hoover vacuum cleaner *Model 150*, 1946

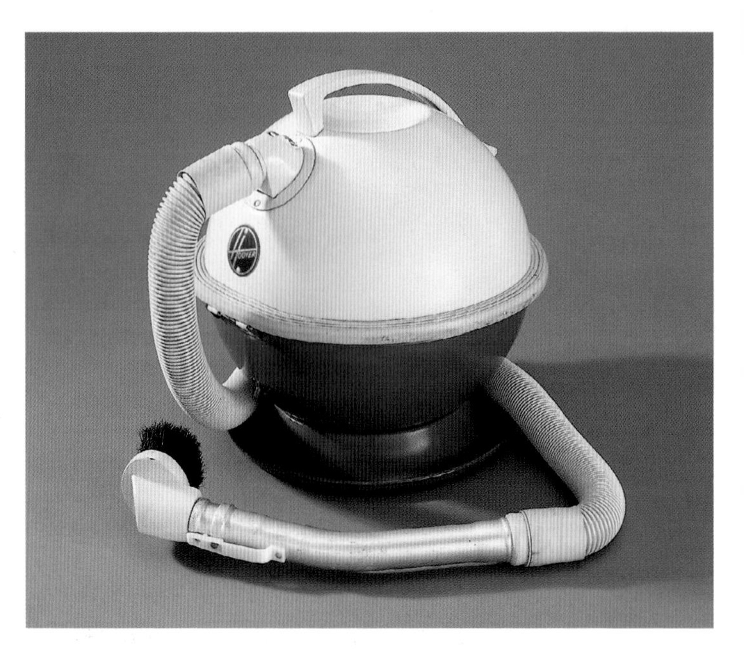

Hoover *Constellation* vacuum cleaner, 1955

Racine, allowed their weight to be decreased to around five or six pounds. The same year, Hoover began a research and development programme, which resulted in the introduction of the "Agitator" to all its upright models in 1926. This innovative feature comprised a rotating bar with spiralled bristles that helped loosen dirt and dust caught in carpets through a beating action. The famous Hoover slogan of 1919, "It beats, as it sweeps, as it cleans" was used very effectively to market this improvement in vacuum cleaner performance. Also in 1926, another feature appeared for the first time – the power switch was moved to the handle on the *Model 700* cleaners. From the 1930s to the early 1950s, the company commissioned **Henry Dreyfuss** to design a range of streamlined cleaners, including one that featured an illuminated lighting strip. In 1955 the futuristic looking *Constellation* cleaner was introduced which, when operating, floated on a cushion of air just like a hovercraft. Extremely popular in America, the design of this model reflected contemporary interest in future space technology. In testimony to the brand's continuing influence, the Hoover name is still recognized around the world, although the Hoover Company is no longer an international company. Maytag Corporation, which acquired Hoover in 1989, divested its European operations (including UK operations) in the mid-1990s.

Eliot Noyes, IBM *Model A* typewriter, 1948

IBM

FOUNDED 1911
NEW YORK, USA

In 1911 three firms – the Tabulating Machine Company (producing electrical machines that processed data using a punched card system), the Computing Scale Company (holding the patent for Julius Pitrat's computing scale) and the International Time Recording Company (manufacturing a mechanical time recorder) – merged and later re-named International Business Machines (IBM) in 1924. Four years later, the data capacity of punched cards almost doubled from 45 to 80 columns of information, which heralded the development in the early 1930s of a new series of machines that could not only perform addition and subtraction, but could also undertake full-scale accounting calculations. In 1935 IBM launched its first electric typewriters and, a year later, provided accounting machines for America's Social Security Program, which was referred to as "the biggest accounting operation of all time". Developed in collaboration with Harvard University, the ASCC (Automatic Sequence Controlled Calculator) was introduced in 1944 and was the first machine capable of automatically executing lengthy computations. Four

Eliot Noyes, 705 *Electronic Data Processing Machine*, 1954

128,00-BIT CMOS SRAM microchip, early 1990s

years later, IBM launched the first computer that combined electronic computation with stored instructions. During the 1940s, **Norman Bel Geddes** advised IBM on the design of its office products. 1952 saw the introduction of the first computer capable of scientific calculations, the IBM *701*, which used magnetic tapes to store data equivalent to 12,500 punched cards. **Eliot Fette Noyes**, who had previously worked in Geddes' office, designed the *705 Electronic Data Processing Machine* in 1954 and two years later was appointed corporate design director of IBM. Noyes was responsible for several revolutionary products, including the *Selectric I* typewriter (1961) with its innovative golf ball typing head. He also developed a strong corporate identity for IBM, both through the design integrity of his own products and by commissioning graphics from Paul Rand and buildings from leading architects such as Marcel Breuer (1902–1981). During the 1950s and 1960s, IBM continued to develop state-of-the-art computing systems for a range of uses, from accounting and ticketing operations to helping NASA with its space programme. Computer processors and peripheral units, such as those of IBM's landmark *System/360* (1964), were still huge in size but were becoming increasingly powerful. In 1968 IBM introduced its *3986* Braille typewriter – an important design for disability – and two years later entered the photocopying market with the IBM *Copier*. During the 1970s, IBM developed the *3614 Automatic Banking Machine* and the *3660 Supermarket System* that could read universal product codes. The late 1970s saw the development of computers that used the new semiconductor technology and electronic typewriters with microprocessors. IBM introduced its first personal computer in 1981, but by the mid-1980s was finding it difficult to compete with **Apple Computer's** more user-friendly models. As one of the largest corporations in America, IBM was slow to react to technological change and as a result began losing market share. As a pioneer in the field of information processing, IBM has nevertheless affected all areas of life – from banking and shopping to energy generation and weather forecasting. The character of the company is hallmarked by technological innovation and a strong design ethos that reflects its core values – respect for the individual, customer service and excellence.

Tracy Currer & Nick Dormon, Digital Radio for BBC, 1997

IDEO

FOUNDED 1969
LONDON, ENGLAND

IDEO's founder, Bill Moggridge (b. 1943), studied industrial design at the Central School of Arts & Crafts, London, graduating in 1965. He later undertook a research fellowship in typography and electronic communications at the Hornsey School of Art. In 1969 he established the design office, Moggridge Associates, and ten years later opened a second office in San Francisco, called I. D.Two, to cater to Silicon Valley's fledging computer industry. One of his first products was the GRiD *Compass* laptop computer (1980). Over the succeeding years, believing that the design of an interface must be an integral part of a product's development, Moggridge christened and pioneered a new discipline known as "Interaction Design". In 1991 Moggridge's office merged with David Kelley Design and was renamed IDEO. By offering a full product development service, the consultancy's interdisciplinary teams translate ideas into ready-to-manufacture products. IDEO's non-linear design methodology, which includes brainstorming sessions, can at times appear chaotic; as the business guru Tom Peters noted: "IDEO is a zoo.... Experts of all flavours commingle in 'offices' that look more like cacophonous kindergarten classrooms than home of one of the world's most successful design firms". This way of working, however, generates a powerful synergy that harnesses both creativity and cutting-edge technology to produce design solutions that are both novel and imaginative. While at the forefront of technological developments, IDEO also realises that emphasis must be placed on human needs, and has therefore adopted a "user-focused' approach to design so as to create user-friendly products. IDEO has designed over 2,000 medical, computer, telecommunications, industrial, furniture and consumer products for a wide-ranging clientele, including **Apple Computers**, **Black & Decker**, Nike, British Airways, Baxter Healthcare, Amtrak, Deutsche Telecom, Microsoft, NEC, Nokia, Samsung, Siemens and Steelcase.

Christopher Loew, *SyncMaster 400 TFT* flat screen monitor for Samsung, 1996

ALEC ISSIGONIS

BORN 1906 SMYRNA (IZMIR), TURKEY
DIED 1988 BIRMINGHAM, ENGLAND

Born in Turkey as the son of a Greek merchant, Alec Issigonis emigrated to England in 1922. Although he apparently received no formal training as an engineer, he did study for a time at Battersea College, London. He was subsequently employed as a draughtsman by Rootes Motors in Coventry and later moved to Morris Motors in Oxford, where he worked as a suspension designer. He later became Morris's chief engineer. 1948 saw the launch of his revolutionary Morris *Minor*, which was influenced by American automotive styling from the 1930s and 1940s. This diminutive yet curvaceous vehicle had a unitary body construction which made it suitable for large-scale mass production. As the first Modern British car, the Morris *Minor* was immensely popular and became the first all-British car to exceed sales of one million. It remained in production until 1971 and to this day has a devoted following of collectors and enthusiasts.
In the 1950s Issigonis briefly worked for other companies before returning to Morris, which had by now been renamed the British Motor Corporation (BMC). In response to petrol shortages resulting from the Suez Crisis and

Advertisement featuring the 1949 model of the Morris *Minor*

the resurgence of interest in the **Volkswagen** *Beetle*, Issigonis designed the small, fuel-efficient and inexpensive Morris *Mini*. Launched in 1959, this front-wheel drive, box-like vehicle was only three metres long. It employed a radical layout that included a space-saving transverse engine, which allowed the compact cab to accommodate four adult passengers in relative comfort. The *Mini* was influential not only in terms of its design but also culturally – it was the first truly "classless" car, as likely to be driven by celebrities as by ordinary citizens. No car came as close to epitomizing the Swinging Sixties as the Mini, although in later years its safety record was seriously questioned. The *Mini* was sold in several versions, including the wooden-framed, van-like *Traveller*. In 1962 Issigonis designed the slightly larger *1100* which was similarly manufactured under Morris, Austin and badges. Like his earlier models, this design was also influential upon later compact cars such as the **Renault** 5 and the Volkswagen *Golf*. In 1967 Issigonis was made a fellow of the Royal Society and two years later was knighted for his immense contribution to the British car industry. He eventually retired in 1971. By 1988 – the year of his death – over five million *Minis* had been sold.

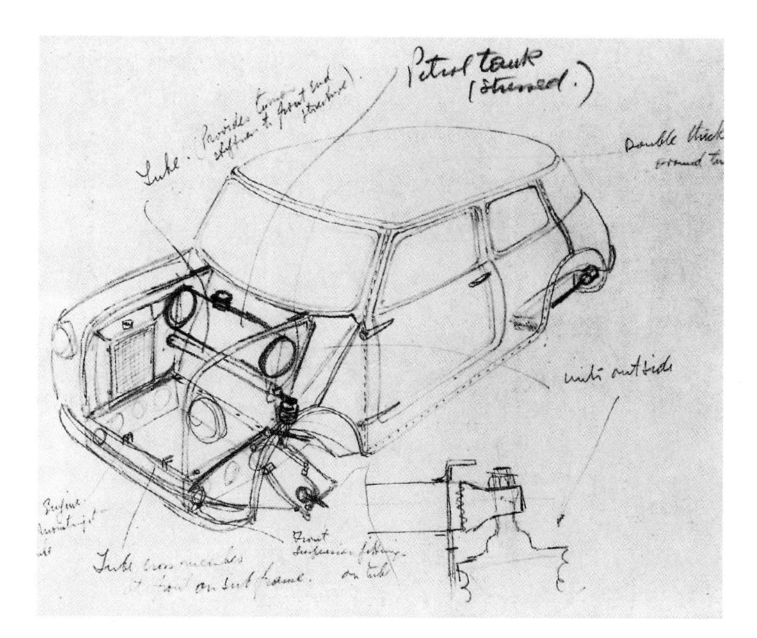

Drawing of the Morris *Mini Minor*. c. 1959

JONATHAN IVE

BORN 1967
LONDON, ENGLAND

Jonathan Ive studied design at Newcastle Polytechnic and later worked for the London-based industrial design consultancy Tangerine, where he designed a wide range of products, from televisions and VCRs to sanitaryware and hair combs. At Tangerine he also assisted in the development of the *PowerBook* for **Apple Computer** in 1991. While working on this project, Ive noted that the bland product identity of computers was the result of their arbitrary configurations, and concluded that there was a huge opportunity in creating exciting new products that did not conform to the conventional grey, beige or black boxes. At this stage, the computer industry was mainly concerned with the internal aspects of its machines – processing speed and memory capacity – and little if any thought was being given to their external form. As a consequence, the industry as a whole was suffering from what Ive describes as "creative bankruptcy". Frustrated that, as an external consultant, he could only have a small impact on the future development of computers, in 1992 Ive joined the design team at Apple, where he became director of design. It was not until Steve Jobs' return to Apple, however, that the design team was given the freedom to concentrate on the "pursuit of nothing other than good design". Jobs realised that Apple needed to recapture its once strong identity, which had been diluted by a "design by focus group" mentality. Significantly, his first day back at the helm marked the beginning of the *iMac* project. The translucent turquoise *iMac* (1998) with its unified curvaceous organic form, broke all conventions. At long last, here was a computer that looked cool and had a strong identifiable character. At its launch, it became obvious that the rather beleaguered Apple (whose market share had shrunk to a miserable 3 % in 1997) had backed a winner that could restore its fortunes – an astonishing 150,000 *iMacs* were sold over the weekend following its introduc-

Advertisement for *iMac*, 1999

iMac, 1998

tion. Helped by a high-profile advertising campaign, which included the insightful slogan "Chic, not Geek", *iMac* became the best-selling computer in America – impressively, because of its design rather than its technology. Ive's designs have managed to powerfully differentiate Macs from PCs, and it may well be that the people who buy into Apple Computer products on the strength of their looks will become hooked on the company's excellent user-friendly operating system.

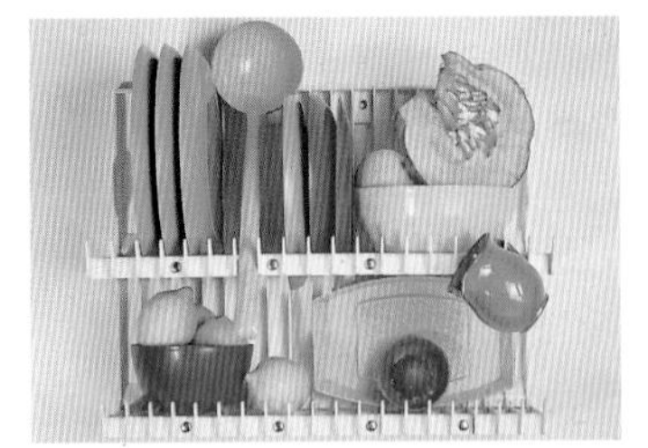

Gino Colombini, *KS 1171* plastic dish rack with plastic dishes, 1954

KARTELL

FOUNDED 1949
MILAN, ITALY

Kartell, a plastics consumer products manufacturing company, was founded by Giulio Castelli (b. 1920) in 1949. Castelli started from the view that: "The public is willing to accept new forms for machines performing new functions, but when century-old objects like a spoon, a chair, etc. are involved, it's not easy to make new aspects accepted. If men fear novelty, let's give them something even more novel." The first product launched by Kartell was an innovative ski-rack designed by the architect and industrial designer, Roberto Menghi (b. 1920). This was followed by a range of household articles, many of which were designed by Gino Colombini (b. 1915). From lemon squeezers and dustpans to washing-up bowls and baby's baths, Kartell transformed every-day objects into sleekly designed polyethylene products that were startlingly forward-looking. The company's design and materials innovations were widely celebrated and its products won numerous Compasso d'Oro awards at Milan Triennale exhibitions. Kartell also produced furniture, including a metal and plastic sectional cupboard system (1956) designed by Gino Colombini and Leonardo Fiori, but it was not until the 1960s that it became widely known in this field,

Anna Castelli Ferreri, *4870* stacking chair, 1987

in particular through its manufacturing of Marco Zanuso (b. 1916) and Richard Sapper's *Model No. 4999/5* stacking child's chair (1961–1964), Joe Colombo's revolutionary *Model No. 4860 Universale* (1965–1967) which was the first adult-sized fully injection-moulded plastic chair, and the injection-moulded ABS *4953–54–55–56* stacking storage cylinders (1970) designed by Anna Castelli Ferrieri (b. 1920). Kartell also produced lamps by Sergio Asti (b. 1926), Marco Zanuso and Achille (1918–2002) and Pier Giacomo Castiglioni (1913–1968). The company maintained a high profile in the 1980s through its production of furniture designed by Philippe Starck (b. 1949), such as the characterful *Dr. Glob* chair (1988). In the 1990s, Kartell's translucent *Mobil* drawer unit (1995), designed by Antonio Citterio (b. 1950) and Glen Oliver Löw and manufactured in injection-moulded PMMA, was awarded a Compasso d'Oro. In 1997 the company began producing Ron Arad's (b. 1951) highly successful *Book Worm* shelving in extruded and injection-moulded technopolymers. Kartell's in-house design studio, Centrokappa, which is directed by Giulio Castelli's wife, Anna Castelli Ferrieri, has also produced some notable designs, including the *5300, 5312, 5320* system of children's furniture. When he founded Kartell in 1949, Castelli stated that he "intended to arrive at the difficult synthesis between technology and design, between economy and the response to social demand". Having successfully struck this balance over its 50-year history, Kartell today can boast that one of its products is sold somewhere in the world every 30 seconds.

Antonio Citterio, *Mobil* drawer system, 1995

KODAK

FOUNDED 1881
ROCHESTER, NEW YORK, USA

George Eastman

At the age of 24, George Eastman (1854–1932) decided to visit Santo Domingo and bought some photographic equipment with which to record his travels. In those days the paraphernalia needed for photography was extensive – an enormous camera and heavy tripod, a dark-room tent and a large assortment of chemicals. In the end Eastman never made the trip, but he became captivated by photography and determined to simplify the process. Having read that British photographers were making and using gelatine emulsions that, when dry, could be exposed at leisure, Eastman made his own using a formula outlined in a magazine. After three years' experimentation, in 1878 he perfected a dry-plate gelatine emulsion. The following year he developed an emulsion-coating machine, which allowed him to begin the

Brownie camera, c. 1902

Advertisement for Kodak cameras, c. 1910

commercial mass production of his dry plates. In 1881 he founded the Eastman Dry Plate Company together with Henry A. Strong (b. 1919). In 1884 the firm introduced Eastman Negative Paper and developed a roll holder for the papers, and in 1885 pioneered the first transparent photographic film. Three years later, the first Kodak camera was launched; costing just $25, it was marketed under the slogan: "You push the button – we do the rest." This revolutionary and easy-to-use camera heralded the advent of amateur photography around the world. In 1889 Eastman marketed the first commercial transparent and flexible film roll, paving the way for the subsequent development of **Thomas Alva Edison**'s motion picture camera (1891). The company was renamed the Eastman Kodak Company in 1892, and in 1895 introduced the *Pocket* Kodak camera, which used roll film. 1900 saw the launch of the famous *Brownie* camera, which was constructed of a pressed cardboard box with a wooden end. Inexpensive and cheap to use, the *Brownie* took perfectly acceptable photographs and made photography accessible to virtually everyone. In 1908 Kodak pioneered the first commercial safety film using a cellulose acetate rather than a flammable cellulose nitrate base. Four years later, it established one of the first industrial research laboratories in America, dedicated to the evolution of new cameras and films. 1923 saw the introduction of 16mm-reversal film, the 16mm *Cine-Kodak* motion picture camera and the *Kodascope* projector, all of which made amateur cinematography possible. 16mm Kodacolour film was launched five years later, together with the first microfilm system for the storage of bank records. Through the 1930s and 1940s, Kodak continued to develop new products, including the first commercially successful colour slide film (1935) and the highly successful *Bantam* camera range, which included the streamlined Kodak *Bantam Special* camera designed by **Walter**

Bantam Colorsnap camera, 1955–1959

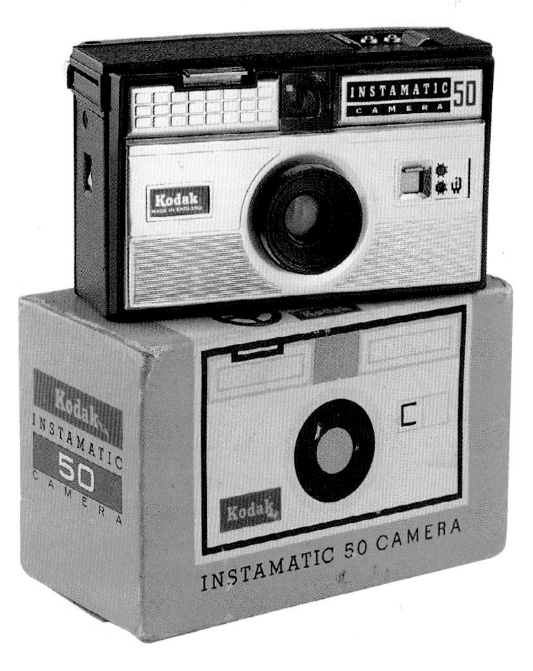

Kodak *Instamatic* 50 camera, 1972

Kenneth Grange, Kodak *Instamatic* 130, 230 and 330 cameras, 1975

Dorwin Teague. Its cameras were often developed by outside design consultants in conjunction with the Kodak Styling Division and the engineering department, and grew increasingly compact and easier to use. Thus the Bantam *Colorsnap* camera of 1954, with its "dial-the-weather" exposure setting, simplified colour photography to such as extent that it was nicknamed "Auntie's Camera". Advances in colour film meant that lightweight snapshot cameras were being manufactured by the early 1960s. Kodak *Instamatic* cameras, which first appeared in 1963, required only very simple exposure controls – just sunny or cloudy flash settings – and featured easy-to-load cartridge film. By 1970, more than 50 million *Instamatics* cameras had been sold, revolutionizing popular photography. By using narrower cartridge film, Kodak further reduced the size of the *Instamatic* so that it could fit inside a pocket. Designed by **Kenneth Grange**, the Kodak *Pocket Instamatic* was extremely popular and over 50 million cameras were sold within seven years. In 1987 Kodak introduced its first one-time-use disposable camera, the Kodak *Fun Saver*, and by 1995 over 50 million of these, too, had been sold. 1995 also saw the launch of the Kodak Digital Camera 40, the first full-feature digital camera to retail for less than $1000. Kodak's historic success has been the result of its "everyman" approach to photography. With film-based photography increasingly becoming a thing of the past, it seems certain that Kodak will continue to be at the forefront of affordable digital camera technology.

Portrait of Raymond Loewy on the cover of *Time* magazine, 1949

RAYMOND LOEWY

BORN 1893 PARIS, FRANCE
DIED 1986 MONACO

Raymond Loewy was the greatest pioneer of industrial design consulting and is still remembered for the humorous adage that summed up his approach to design – "Never leave well enough alone". At the age of 15, Loewy designed, built and flew a toy model aeroplane that won the then-famous James Gordon Bennett Cup. Around this time, he also designed and patented a model plane powered by rubber bands. Having subsequently sold the rights to this toy, called the *Monoplan Ayrel*, to a company that marketed it across France, Loewy learnt that "design could be fun and profitable". With the money he earned from this venture, Loewy was able to study at the Université de Paris and later at the École de Lanneau, from where he received an engineering degree in 1918. During the First World War he served in the French army as a second lieutenant. After his demobilization, he travelled

First redesign of the Gestetner duplicator machine, 1929

Coldspot refrigerator for Sears, 1934

to America in 1919. Arriving in New York, he was utterly amazed by what he later described as "the chasm between the excellent quality of American production and its gross appearance, clumsiness, bulk and noise". He initially found work as a window dresser for Macy's, Saks Fifth Avenue and Bonvit Teller. From around 1923 until 1928, he enjoyed a "pleasant but superficial career" as a fashion illustrator for *Vogue*, *Harper's Bazaar* and *Vanity Fair* among others. In 1923 he also designed the trademark for the Neiman Marcus department store. Loewy eventually left the fashion world to set up his own industrial design office in New York in 1929. Always a self-promoter, Loewy had a card printed that read: "Between two products equal in price, function and quality, the better looking will outsell the other", and sent it to everyone he knew. Soon afterwards he received his first brief, namely to redesign the casing of Sigmund Gestetner's duplicator machine. For this project, he used modelling clay to create a sleek form – a technique he later employed to great effect for his automotive designs. The resulting machine not only looked better but, thanks to its simplified form, was also easier to manufacture and maintain than earlier models. In 1932 Loewy designed his first car, the *Hupmobile*, which was less boxy than existing automobiles. His improved, tapering model of 1934 featured innovative integrated headlamps and predicted the streamlined forms for which he later became so famous. That same year, Loewy also designed the *Coldspot* refrigerator for Sears Roebuck, which was the first domestic appliance to be marketed on the

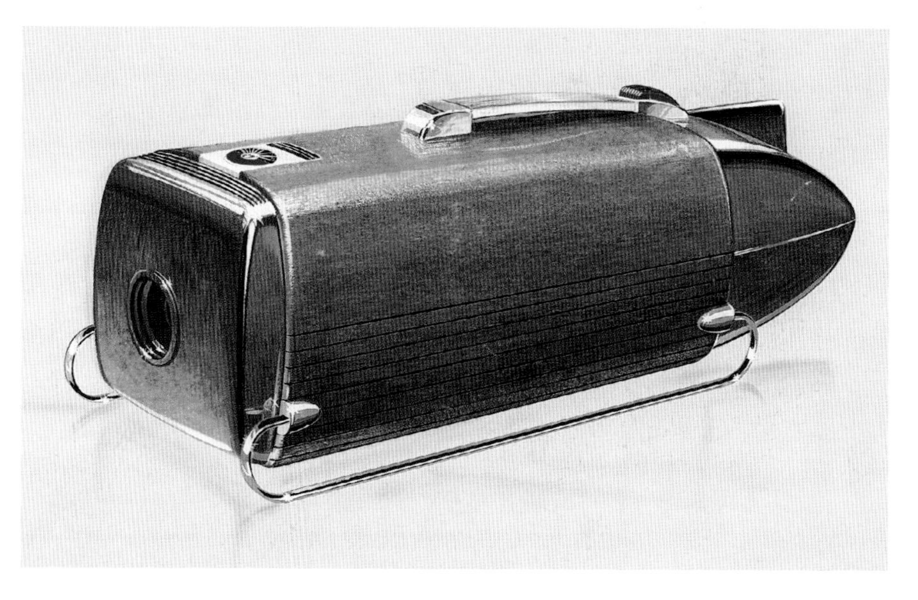

Design for vacuum cleaner for Electrolux, 1939 – this model was never put into production

↑Studebaker, 1939

↗Studebaker, 1950

↑Studebaker *Starliner*, 1953

↗Studebaker *Avanti*, 1961

strength of its aesthetic appeal – contemporary advertisements invited consumers to "Study its Beauty". The *Coldspot* was also alluring to customers because its improved design had reduced its cost of manufacture, and this was reflected in its competitive retail price.

Also in 1934, the Metropolitan Museum of Art in New York displayed a mock-up of Loewy's glamorous design office, thus illuminating his rise in status from product designer to celebrity industrial design consultant. From 1935, Loewy was commissioned to re-plan several large department stores, including Saks Fifth Avenue. During this period he was also designing aerodynamic locomotives, such as the *K4S* (1934), the *GG-1* (1934) and the *T-1* (1937), and in 1937 published a book entitled *The New Vision Locomotive*. In 1946 he famously remodelled coaches for Greyhound and in 1947 designed his innovative *Champion* car for Studebaker – a precursor of his later European-styled *Avanti* car, also for Studebaker. Not only could streamlining make transportation look more appealing, it also often improved performance and efficiency because of the effect of aerodynamics. Loewy regarded the design of a casing or sheath as an opportunity to allow a machine or a product to express itself. Unlike so many Modernists who allowed form to be completely dictated by function, Loewy balanced engineering criteria with aesthetic concerns in order to achieve what he believed to be the optimal solution. Loewy also became celebrated for his corporate identity work, and in particular for his repackaging of Lucky Strike cigarettes (1942). His presti-

Greyhound *Scenicruiser* bus, 1954

Interior of a Greyhound bus, c. 1946

Logo for Exxon, 1966

→Logo for Shell, 1967

Redesign of Lucky Strike packaging for the American Tobacco Company, 1942

→Redesign of Coca-Cola dispenser *Dole Deluxe*, 1947

gious clientele included Coca-Cola, Pepsodent, the National Biscuit Company, British Petroleum, Exxon and Shell. By 1939 his design office had branches in Chicago, São Paulo, South Bend and London. In 1944 he founded Raymond Loewy Associates with four other designers. In 1949 Loewy became the first designer to be featured on the front cover of *Time* magazine; his picture accompanied by the memorable caption: "He streamlines the sales curve." That same year, Loewy expanded his operations and founded the Raymond Loewy Corporation to undertake architectural projects. During the 1960s and 1970s, Loewy worked as a design consultant to the United States Government, most famously redesigning Air Force One for John F. Kennedy and designing interiors for NASA's *Skylab* (1969–1972). Always putting the consumer first, Loewy's MAYA (Most Advanced, Yet Acceptable) design credo was crucial to the success of his products. Undoubtedly the greatest pioneer of streamlining in the 20th century, Raymond Loewy clearly demonstrated that the success of a product depends as much on aesthetics as it does on function. Few design consultants have been as influential or prolific as Loewy, nor as misunderstood; for although he was a styling genius, he also skilfully improved the design of many products and pioneered numerous design innovations. Signficantly, Raymond Loewy glamorized the practice of design and in so doing raised the status of industrial design as a whole.

DIETRICH LUBS

BORN 1938
BERLIN, GERMANY

Dietrich Lubs studied shipbuilding and later trained as a designer at **Braun**. In 1962 he began working for the company's design department, headed by Dieter Rams (b. 1932). From 1971 he has been responsible for product graphics and has also been active as a product designer. He has designed numerous pocket calculators, alarm clocks and watches for Braun, all of which are distinguished by geometric forms and functional layouts. The majority of his clock designs have matt black plastic bodies, which allow the white and yellow numerals and hands to stand out so that they can be easily read. Since 1995 Lubs has been deputy director of the Braun Design Department. His many product designs are typical of the rational language of design pioneered by Rams at Braun, while also being characteristic of the attributes generally associated with German industrial design – functionalism, logic and high quality engineering.

Pocket calculator, variant of *ET 44* co-designed with Dieter Rams, 1978

Yoichi Takahashi, *MS4/M40* video camera for Panasonic, 1995

MATSUSHITA

FOUNDED 1918
OSAKA, JAPAN

Konosuke Matsushita (1894–1989) founded the Matsushita Electric Housewares Manufacturing Works to produce an adapter socket that he had designed. He later developed an improved two-way socket that was better than any other on the market, and was soon flooded with orders. In 1922 he built a factory in Osaka and a year later developed a revolutionary bullet-shaped, battery-powered bicycle lamp that was marketed under the "National" name. In 1931 his factory also started manufacturing radio and dry cell batteries. Konosuke Matsushita was an industrialist with a social mission who believed that mass production would contribute to "the growth of human civilization". In 1935 the company began researching television technology and was incorporated as the Matsushita Electric Industrial Co. During the war, MEI constructed ships and aeroplanes for the Japanese military and was subsequently branded a *zaibatsu* company by the Allied powers. By 1950, however, such restrictions had been lifted and in 1951 Konosuke Matsushita travelled to America for the first time. MEI produced its first washing machine that same year, followed by its first black-and-white television (1952), refrigerator (1953), tape recorder (1958), colour television (1960) and microwave oven (1963). Today, Matsushita is the world's largest manufacturer of consumer electrical products, which it markets under the brand names of Panasonic, Quasar, Technics, Victor and JVC. The Matsushita concept of product design "incorporates both function and beauty, maximizing the interface between user and machine to enhance the comfort and convenience of people's lives". The company's basic principles of product design are: fun and ease of use, innovation, universality, environmental responsibility and facilitative of new lifestyles.

Model R-72S wrist-radios, 1969

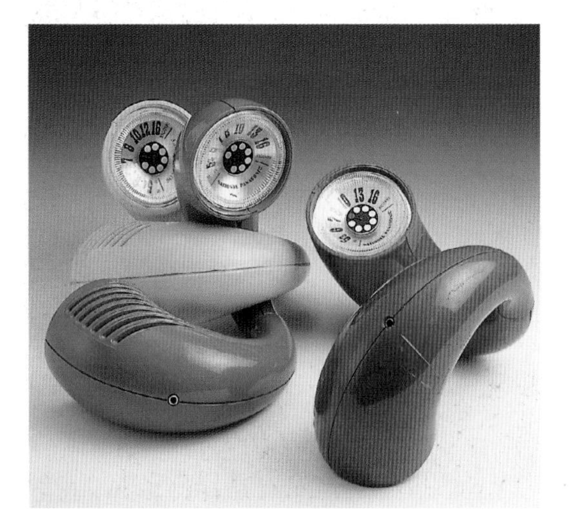

Douglas *DC-3*, 1935

MCDONNELL DOUGLAS

FOUNDED 1967
ST. LOUIS, MISSOURI, USA

The McDonnell Douglas Corporation was formed in 1967 as a result of a merger between the McDonnell Aircraft Corporation, founded by James S. McDonnell (1899–1980) in St. Louis in 1939, and the Douglas Aircraft Company, founded by Donald W. Douglas (1892–1981) in Santa Monica in 1920. As a civil engineering assistant at the Massachusetts Institute of Technology's department of aerodynamics, Douglas helped develop one of the first wind tunnels for the testing of aircraft (1914–1915). In 1920 he founded his company and the same year designed the *Cloudster*, the first aerodynamically streamlined aeroplane and the first aeroplane able to carry a load that exceeded its own weight. Recognizing the passenger and cargo-carrying potential of aeroplanes over long distances, in 1932 Douglas began developing his famous *DC* (Douglas-Commercial) series of aircraft. This included the immortal *DC-3* (1935), which was the world's first successful commercial airliner. Powered by twin Pratt & Whitney engines, it could carry 21 passengers at a speed of 195 mph.

Douglas *DC-3* operated by KLM airlines, late 1930s

During the Second World War, the *DC-3* was converted for military use and re-designated the *C-47*. In total, the Douglas Aircraft Company supplied 29,000 warplanes to the US military (including over 10,000 *C-47s*) – one-sixth of the airborne fleet. After the war, the company continued developing commercial passenger aircraft, including the advanced piston-engined *DC-7*, whose range made non-stop coast-to-coast services possible for the first time. With the advent of passenger jet aircraft, Douglas developed the *DC-8* (1958) and *DC-9* (1965), but nevertheless found itself lagging behind **Boeing**. As a result of this, the merger with McDonnell was sought.

Having been founded on the eve of the Second World War, the McDonnell Aircraft Corporation grew quickly and became a major US defence supplier. During the war it began research into jet propulsion, which led to the development of *FH-1 Phantom* (1946), the world's first operational carrier-based jet fighter. As the first US Navy aircraft able to attain a speed of 500 mph, the *FH-1* was the progenitor of a long and distinguished line of McDonnell fighter aircraft, including the *A-4 Skyhawk* (1954), the super-versatile Mach 2 *F-4 Phantom* (1958), the *F-15 Eagle* (1972) – still the US Air Force's premier fighter – and the *F/A-18 Hornet* (1978). Both McDonnell and Douglas were at the forefront of Space Age technology and made important contributions to the Mercury, Gemini and Saturn/Apollo space programmes. Today, McDonnell Douglas' commercial transports, such as the impressive *MD-11* wide-body tri-jet (1990), combat aircraft and space vehicles add new lustre to both companies' long history of aerospace accomplishments. In 1997 the McDonnell Douglas Corporation merged with Boeing.

F-4A Phantom, 1961

F-15C Eagle, 1978

F-15E Eagle, 1989

MERCEDES-BENZ

FOUNDED 1926
STUTTGART, GERMANY

Benz poster, 1910

Mercedes-Benz is the luxury car brand of the Daimler-Benz company, which was formed in 1926 through the merger of two pioneering German automobile companies, Benz & Co founded by Karl Benz (1844–1929) in 1883, and the Daimler-Motoren-Gesellschaft founded by Gottlieb Daimler (1834–1900) in 1890. Karl Benz initially established his company to manufacture stationary internal-combustion engines. In 1886, however, he launched the three-wheeled *Motorwagen*, which was the world's first practical car powered by an internal-combustion engine, and patented it the following year. Meanwhile, Gottlieb Daimler and Wilhelm Maybach (1846–1929) had also been developing internal-combustion engines from 1883 onwards, and in 1885 they patented a four-stroke version. That same year, they mounted one of their engines onto a wooden-framed bicycle and thus created the world's first motorcycle. In 1886 they motorized a previously horse-drawn carriage; this was a revolutionary design in that it was propelled by an "invisible" power source. In 1890 Daimler finally founded the Daimler-Motoren-Gesellschaft. In 1894

Mercedes car, 1904

Mercedes-Benz *260 Stuttgart* roadster, 1928

Mercedes-Benz *150 H*, 1934

Mercedes-Benz *190 SL*, 1955

Mercedes-Benz
500 SL, 1991

the company launched a belt-driven car, featuring a four-speed belt transmission connecting the engine to the back wheels, which allowed smoother gear changing. Benz, meanwhile, had its own new car to launch that year, the *Benz Velo*, which was the world's first series production car. By 1899 it had developed a racing car, one of the finest and most reliable of its time. Daimler followed this in 1906 with its first Mercedes racing car. In 1926 the two companies merged to create the Daimler-Benz company, which continued to produce cars under the Mercedes-Benz name. These included the famous *SSK* sports car (1928) and the *260 Stuttgart* (1928), which was the company's mainstay during the late 1920s. During the 1930s, as well as producing elegantly styled saloons and sport cars such as the *159 H* roadster (1935), the company also designed and built aerodynamic racing cars such as the *W 25* (1934) and the *W 125* (1937), the latter scoring 27 victories. During the Second World War, Daimler-Benz's factories were turned over to military production. Car production resumed in 1946, but it was not until 1950 that the company launched a new model, the *170 S* cabriolet, which became a powerful symbol of Germany's economic recovery. 1955 saw the introduction of the *180* – the first Mercedes with a modern three-box body – and the elegant *190 SL*, which was described as "the ideal vehicle for sporting ladies". Over the succeeding decades, Daimler-Benz continued to produce cars, vans, trucks and buses and by 1965 it was the largest manufacturer of commercial vehicles in Europe. In 1988 Mercedes-Benz produced its ten millionth car. Having merged with the **Chrysler** Corporation in 1997 to become DaimlerChrysler, the company continues its long tradition of producing high-quality innovative vehicles – from the top-of-the-range *S-Class* to the compact *A-Class* – that exemplify German design at its best.

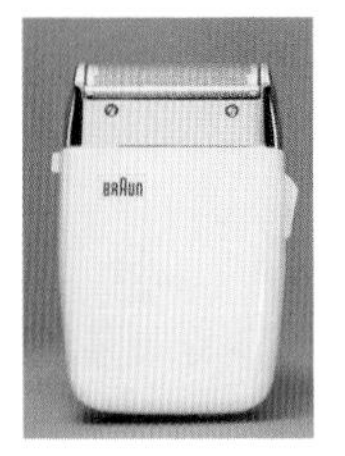

SM 3 electric shaver, 1960 (co-designed with Hans Gugelot)

GERD ALFRED MÜLLER

BORN 1932 FRANKFURT/MAIN, GERMANY
DIED 1991 ESCHBORN, GERMANY

Gerd Alfred Müller apprenticed as a joiner prior to studying interior design at the Werkkunstschule in Wiesbaden. Between 1955 and 1960 he worked in the prestigious design department at **Braun** headed by Dieter Rams (b. 1932). While there, he designed many of the company's best-known household appliances and electric shavers. Typical of his work for Braun are his *Blender MX 3/MX 32* liquidizer (1958/1962) and *Multimix KM 3/KM 32* food processor (1957/1964), whose designs are distinguished by a remarkable clarity and underlying logic that made them easy to use and at the same time aesthetically pleasing. The neutral and rational styling of these products exemplified Braun's essentialist approach to design, which was founded on Rams' belief that it is the responsibility of the designer to create order in a world of visual noise and that products should therefore be as unobtrusive as possible. In 1960 Müller established his own office in Eschborn and began working as a freelance industrial and graphic designer. The solidity and purity of Müller's products were highly influential upon the design of succeeding generations of household appliances.

Multimix KM 3 food processor for Braun, 1957

Nissan assembly line, 1990s

NISSAN

FOUNDED 1934
YOKOHAMA, JAPAN

After the launch of the original *Dat* car by the Kwaishinsha Company in 1914, in 1931 the Tobata Casting Co. began producing a new generation of cars which it christened Datsun ("son of Dat"). Two years later, another company, founded by Yoshisuke Aikawa (1880–1967), took over their manufacture. In 1934 this new venture was renamed Nissan Motor Company and its founder, who had a grand scheme of mass-producing 10,000–15,000 cars per annum, began putting his plan into action. The first small-sized Datsun rolled off the assembly line at Nissan's plant in Yokohama just one year later, symbolizing Japan's rapid pre-war industrialization. During the Second World War, production switched to military trucks and engines for aeroplanes and torpedo boats. Nissan resumed production of non-military vehicles in 1945 and two years later recommenced manufacturing Datsun cars. In 1958 Nissan began exporting Datsun cars to the United States and sold its first Datsun compact pick-up truck there a year later. During the 1960s and early 1970s, Nissan continued to make inroads into the American automotive market, with annual sales rising to 255,000 cars in 1971. The oil crisis massively boosted American sales as consumers began opting for smaller, less expensive and more fuel-efficient automobiles such as those exported by Nissan, Honda and Toyota. By 1973, one million Datsun vehicles had sold in the United States, and two years later the company became the top vehicle exporter to North America. During the 1980s, Nissan established two strategic manufacturing bases in America and Britain and in 1983 began marketing cars under the Nissan name, such as the hugely successful Nissan *Sunny* launched in 1985. By 1987 its cumulative exports had exceeded 20 million units. In the early 1990s, Nissan became one of the first automotive manufacturers to adopt soft design with its introduction of the *Micra* (1992), which was named Car of the Year in Europe in 1993.

Nissan *Micra*, 1992

ELIOT FETTE NOYES

BORN 1910 BOSTON, MASSACHUSETTS, USA
DIED 1977 NEW CANAAN, CONNECTICUT, USA

Eliot Fette Noyes studied architecture at Harvard University and at the Harvard Graduate School of Design. In 1939 he joined Walter Gropius and Marcel Breuer's architectural practice. Upon Gropius' recommendation, he was appointed the first director of industrial design at the Museum of Modern Art, New York – a position he held from 1940 to 1942 and 1945 to 1946. Between 1946 and 1947 Noyes was design director of **Norman Bel Geddes'** industrial design practice, which consulted to **IBM**, and in 1947 established his own office in New Canaan, Connecticut. From 1956 to 1977 he was corporate design director of IBM, during which time he designed several revolutionary products, most notably the *Selectric I* golf-ball typewriter (1961). Noyes helped establish a strong corporate identity for the company not only through the integrity of his own product designs, but by commissioning graphics from Paul Rand (1914–1996) and buildings from Breuer. Noyes worked as a design consultant to many other companies, including Westinghouse, Mobil, **Xerox** and Pan Am. As one of the most influential advocates of Good Design, Noyes reshaped entire corporations and set new standards for product design in America.

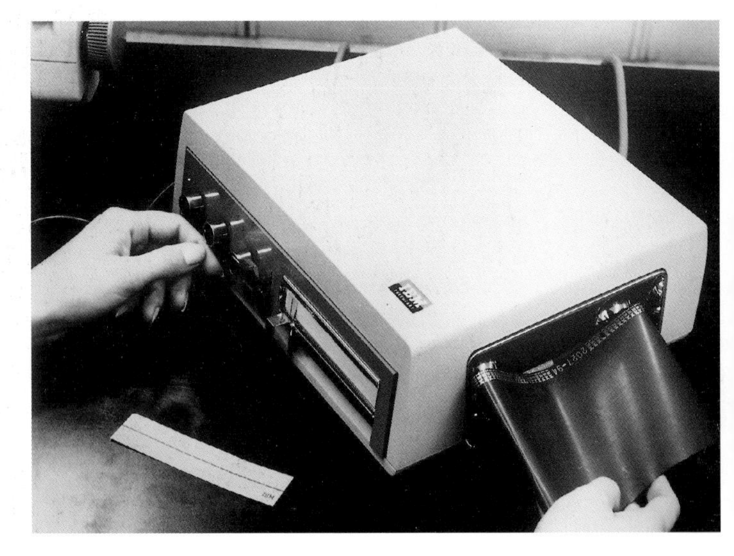

Executary secretarial transcribing machine, *Model 212*, for IBM, 1961

Selectric 1 golf-ball typewriter for IBM, 1961

Detail showing golf-ball element of the *Selectric 1*, 1961

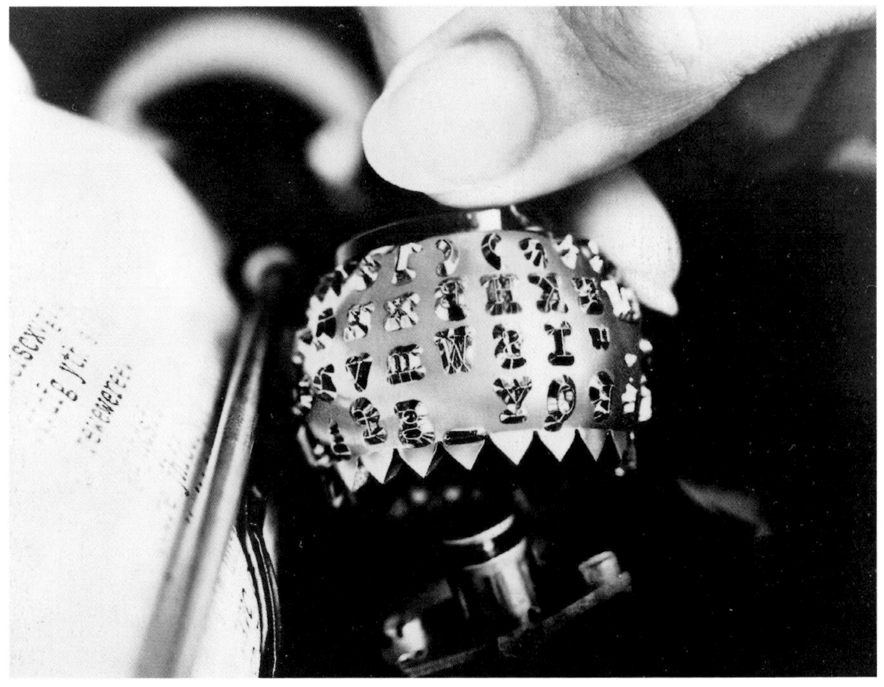

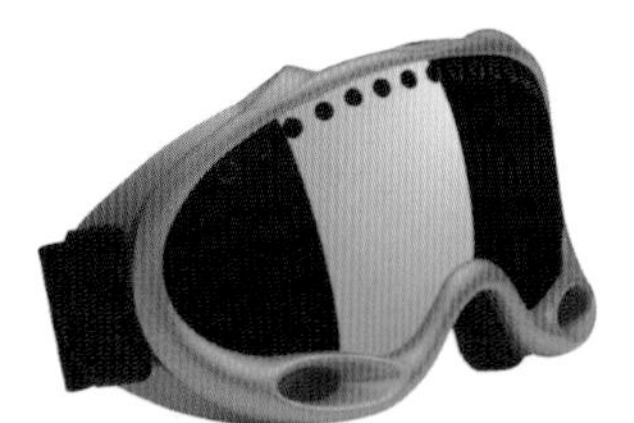

OAKLEY

FOUNDED 1975
FOOTHILL RANCH, CALIFORNIA, USA

A Frame snow goggle, 1999

Founded by Jim Jannard in 1975, Oakley has a worldwide reputation for the superlative design of its eyewear products. Since its inception, Oakley has defied convention while focusing on new designs, new materials and new production techniques. In 1984 Oakley came up with the concept of "sculptural physics" and subsequently introduced designs, such as the *XYZ Optics*, that exemplified style and performance. Designs such as these completely redefined the eyewear market, which had been stultified by a handful of major manufacturers who believed they "knew it all". According to Oakley, "an idea is born in the depths of our design bunker, given form with CAD-CAM modelling, brought to life with SLA liquid laser prototyping, tested with spectrophotometers, environmental simulation chambers and ANSI Z87.1 impact rigs" before undergoing rigorous field testing by athletes for whom the designs are originally intended. The form of Oakley's eyewear is dictated by performance: the *Racing Jacket* (described as "chiselled adrenaline"), for example, has a vented and hingeless frame that maximizes peripheral vision and provides de-fogging ventilation, while its "nosebomb" and "earsock" features ensure that the lightweight frame remains securely in place. Oakley has managed to position itself as a major global brand and as well as its innovative eyewear – from sunglasses to snow goggles – it also manufactures sports apparel, footwear and wristwatches.

Racing Jacket eyewear, 1999

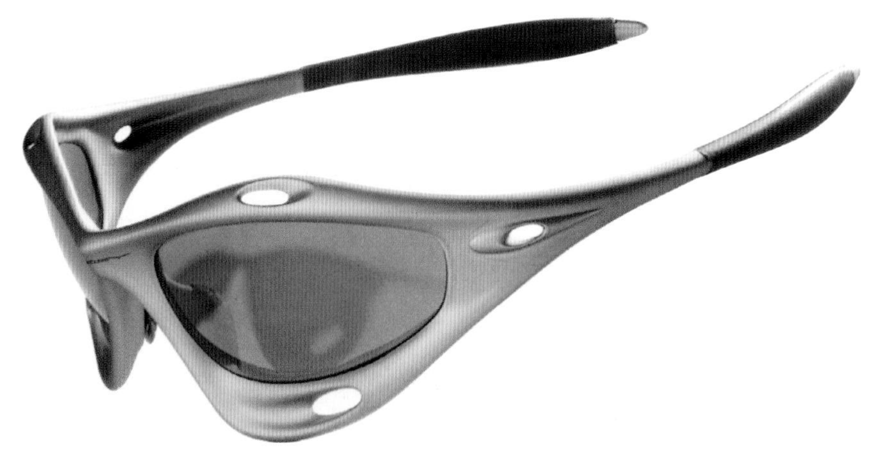

OLIVETTI

FOUNDED 1908
IVREA, ITALY

Camillo Olivetti

Born in Ivrea, Samuel David Camillo Olivetti (1868–1943) studied engineering at the Politecnico di Torino under Galileo Ferraris, who was the discoverer of rotating magnetic fields. After graduating, Olivetti went to London to continue his studies. In 1893 he accompanied Ferraris to the Electricity Congress held in Chicago, and later took a physics course at Stanford University, where he also worked as an assistant electrical engineer. After returning to Italy in 1894, he established a small factory by the name of C. G. S. (Centimetre, Gram, Second) to manufacture electric measuring instruments, which he moved to Milan in 1903. Four years later, he decided to move back to Ivrea and subsequently established Italy's first typewriter factory there in 1908. Aware of American mass production methods, Olivetti began industrially manufacturing typewriters, and in 1909 Ing. C. Olivetti & C. SpA introduced its first typewriter, the *M1*, which was described by a contemporary critic as "robust and elegant" and was praised for its faster carriage speed and smoother key movements. The company grew rapidly, and in the 1920s Camillo's son Adriano Olivetti (1901–1960) was sent to the United States to observe first-hand the latest mass-production techniques, with a view to adopting them back in Ivrea. In 1931 Olivetti launched the *M40* typewriter and the following year unveiled its first portable model, the

Camillo Olivetti, *M1* typewriter, 1910–1911

MP1. As a moral crusader and social reformer, Camillo Olivetti established a foundation in 1932 that offered his workers social security benefits on a level unprecedented in Italy. Other far-sighted employee incentives were later provided, including seaside summer camps for the workers' children, a cafeteria, a kindergarten, a library and housing. In 1933 Adriano Olivetti was appointed managing director and began diversifying the company's product line, launching its first teleprinter in 1937 and its first adding machine in 1941. He also raised the profile of the company by developing a strong corporate identity through product design, architecture, exhibition design, advertising and graphics. Indeed, during the immediate post-war period, Olivetti was one of only a handful of companies worldwide with a truly Modern image. As a consequence, it was also exceptionally influential upon the design of other firms' corporate identities. As an "intelligence co-ordinator", Adriano Olivetti commissioned leading designers to produce cutting-edge product designs, such as the sculptural *Lettera 22* typewriter (1950) by Marcello Nizzoli (1887–1969). He also commissioned leading

Marcello Nizzoli, *Lexikon 80* typewriter, 1948

Ettore Sottsass & Hans von Klier, *Editor 4* typewriter, 1964–1969

Mario Bellini, *Quaderno* lap-top computer, 1989

Michele De Lucchi, *M4–82 Modulo* computer, 1993

Michele De Lucchi, *OFX 500* phone/ facsimile machine, 1998

graphic designers, such as Giovanni Pintori (b. 1912), to produce eye-catching posters and advertisements. Although Olivetti introduced Italy's first electronic computer in 1959 – the *Elea 9003* designed by Ettore Sottsass (b. 1917) – the company was obliged for financial reasons to sell its Electronics Division after the death of Adriano Olivetti in 1960. The company continued its research into electronic processing systems, however, and in 1965 it launched the *P101* programmable desktop computer, an innovative forerunner of the personal computer. During the late 1960s and early 1970s the company launched other notable products, including the *Editor 4* typewriter (1964–1969) designed by Ettore Sottsass and Hans von Klier (1934–2000) and the *Divisumma 18* calculator designed by Mario Bellini (1973). In 1969 Olivetti also introduced the bright red *Valentine* portable typewriter designed by Ettore Sottsass and Perry A. King. A year later, the Olivetti trademark, which is still in use today, was designed by the famous graphic designer Walter Ballmer (b. 1923). While Olivetti experienced financial difficulties in the late 1970s, it nonetheless developed a number of key products, including the company's first electronic typewriter in 1978 and its first personal computer in 1982. During the 1980s Olivetti expanded its operations in the IT field and in the 1990s concentrated on its telecommunications activities. Today, the Olivetti Group is made up of twelve companies operating in either the information technology or telecommunications fields. For Olivetti, "industrial design is a range of activities that not only creates a visual image, but above all contributes to the development of the project as a whole. The designer, in other words, is no longer regarded simply as an expert in aesthetics or style, but as a specialist in the relationship between man and machine."

Workers putting the finishing touches to the newly launched *Model 444* broadcast receiver (soon afterwards renamed the *People's Set*) at the Philco radio factory in Perivale, 1936

PHILCO

FOUNDED 1892
PHILADELPHIA, PENNSYLVANIA, USA

Philco was established as the Helios Electric Co. and initially produced batteries. In 1909 it was re-named the Philadelphia Battery Storage Company (Philco) and in 1927 began manufacturing radios. As early as 1929, it was mass-producing radios using assembly-line techniques and rapidly became one of the "big three" radio manufacturers. In 1932 Philco made inroads into the British market by establishing a custom-built manufacturing facility in Perivale. In 1936 Lord Selsden's Ullswater Committee investigated the high price of radios in Britain and challenged the whole industry to produce better and cheaper models. That same year, Philco responded by launching its one-piece Bakelite *Model 444* radio, which sold for just six guineas. Like Walter Maria Kersting's earlier *Volksempfänger*, the radio had a standardized design that was specifically intended for mass production. Initially it did not sell well, but when it was renamed the *People's Set* it became extremely popular and over 500,000 units were sold. During the 1930s, Philco also financed the researches of the American Philo Taylor Farnsworth (1906–1971), an early pioneer of television. From the late 1940s to the mid-1950s, the company produced a wide range of technically sophisticated televisions housed in historicist cabinets. In the late 1950s, however, Philco began using futuristic styling for its television sets. Known as the Philco *Predictas*, these Space-Age televisions were the most distinctive sets ever designed in America. Despite the use of highly innovative forms and features such as swivelling picture screens, Philco was dogged by the poor picture quality of its televisions and eventually went out of business in 1962.

Philco, *Model 444* broadcast receiver, 1936

Philips *All Transitor* reel-to-reel tape recorder, c. 1960

PHILIPS

FOUNDED 1891
EINDHOVEN, THE NETHERLANDS

In 1891 Gerard Philips founded a company to "manufacture incandescent lamps and other electric products". The firm concentrated on the production of carbon-filament lamps and by 1900 had become one of the largest manufacturers of its kind in Europe. As a means of stimulating product innovation, in 1914 Philips established a research department that explored both physical and chemical phenomena. In 1918 the company manufactured its first X-ray tube and from 1925 began conducting experiments into television. Two years later, Philips launched its first radio and by 1932 it had sold over one million. The following year, the company produced its 100 millionth radio valve and began manufacturing X-ray equipment in the United States. Around 1950, having previously pursued an only loose approach to design, Philips' appointed the architect Rein Veersema to oversee the design of the company's shavers, radios, televisions and record players. During his 14-year tenure at Philips, Veersema instituted a systematic design strategy that addressed every aspect of the design process, from ergonomics research to product costings. He also promoted the idea of developing product families so as to project a coherent corporate identity. Many of the designers employed by Philips had worked previously in the **Braun** design department and were thus particularly adept at designing products within the parameters of a house style. During this period, Philips became a leading pioneer of sound recording technology and in 1963 introduced its landmark Compact Audio Cassette, which was soon after adopted as an international standard format. In 1972 the company had also developed a laser videodisc, but by the time this was finally introduced in 1978 the market was already saturated with video tape systems. In the course of developing the videodisc, however, it was discovered that it offered significantly better

Philips Argenta/TSF Brevets Français Philips enamelled metal sign, c. 1920s

Philips Corporate Design, *Sensuval* television, 1995

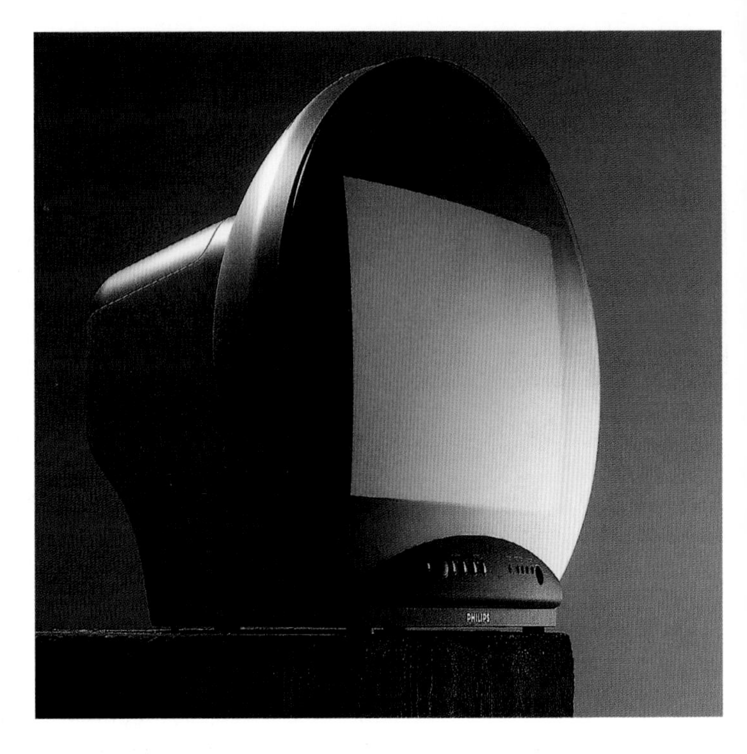

sound reproduction than conventional vinyl records. This led directly to the introduction of Philips' revolutionary compact disc (CD) in 1983, which in a stroke rendered the gramophone record virtually obsolete and seriously threatened the future existence of audio cassettes. From 1980 Philips' design team was headed by the American industrial designer and architect Robert Blaich, who espoused the concept of "global design" and championed "product semantics" in an attempt to compete with Japanese companies. Since then, Philips has continued to strengthen the visual identity of its products with an increasingly "high-tech" aesthetic. Through its commitment to design, Philips has developed an exceptionally strong brand that is guided by its corporate motto: "Let's Make Things Better".

Vespa motor scooters launched by Piaggio in 1946

PIAGGIO

FOUNDED 1884
SESTRI PONENTE, ITALY

Rinaldo Piaggio (1864–1938) founded the Società Rinaldo Piaggio in 1884. Subsequently renamed Piaggio & C., the company initially specialized in luxury ship fitting but eventually began manufacturing railway carriages, engines, coaches, vans, trams and truck bodies. During the First World War, production included aeroplanes and seaplanes, and the company later manufactured several notable aircraft, including the futuristic-looking *P7* seaplane (1929) and the all-metal *P16* bomber (1935). Piaggio's plants were virtually all destroyed in the Second World War, but the company decided to rebuild and address the pressing need for an inexpensive means of transportation. This led to the development of a scooter, which was designed by the aircraft engineer Corradino D'Ascanio (1891–1981) and whose now famous name was coined by Enrico Piaggio, who when he first saw the new design exclaimed: "It looks just like a wasp" (wasp = *vespa*). After being patented, the *Vespa* went into production and became an immediate success. By 1956, one million *Vespas* had been produced and the scooter became synonymous with the Italian Reconstruction and Latin high-life. Today, the *Vespa* is enjoying an unprecedented revival of popularity as a stylish, practical and affordable form of transport.

Corradino d'Ascanio, *Vespa 125* motor scooter, 1951 – original version designed in c. 1945

Professor Ferdinand Porsche

PORSCHE

FOUNDED 1931
STUTTGART, GERMANY

The design practice that would later evolve into Porsche AG was founded in Stuttgart in 1931 by Ferdinand Porsche (1875–1951), who went on to design the hugely successful **Volkswagen** *Beetle* (1934–1938). During the Second World War, Porsche designed several military vehicles, including the *Tiger* tank. After the war, his son "Ferry" used Volkswagen components and a flat four-cylinder engine to create the *356* roadster (1948), which became the first vehicle to carry the Porsche name. The streamlined body of this sports car was designed by Erwin Komenda (1904–1966} not only to look beautiful, but to be as functionally efficient as possible. By 1958, over 10,000 *356*s had been manufactured. In 1961 the company began work on a new model with a rear-mounted, air-cooled six-cylinder engine and a body designed by Komeda and Ferry Porsche's son **Ferdinand Alexander Porsche**. This resulted in the immortal *911*, which was first produced in 1964 and went on to become one of the most famous sports cars of all time. Still in production today, numerous evolutions of the *911* were manufactured, including the *911 Carrera RS* (1972), which was the first production car with a front and rear spoiler, and the *911 Turbo* (1974), which was the world's first production car with an exhaust gas turbocharger. Porsche's first front engine transaxle configured car, the *924*, was introduced in 1976 and was followed by the similarly configured *928* in 1977, which was conceived – controversially to purists – as the successor to the *911*. By the mid-1990s, however,

Evolution of the Porsche *911 Carrera*, 1972–1998

both these models had been discontinued, despite healthy sales. As well as producing highly acclaimed road cars, such as the limited edition *959* supercar (1985), Porsche has also produced numerous racing cars, including the *917* (1970) and the *956* (1982) – two of the most successful racing cars ever. In 1993 Porsche launched a new concept car at the Detroit Auto Show – the six-cylinder mid-engine *Boxster*. Production of this roadster began in 1996 and it became an immediate success. Through its commitment to design, engineering and technological excellence, Porsche has established one of the strongest brands in Germany and its name remains synonymous worldwide with the sports car. In the company's own words: "For Porsche, the ultimate test is to exhaust all possibilities for excellence even at the cutting edge of technology. We regard every idea as an opportunity. It is this perception that has created our company and what guides us still. It is the very essence of Porsche".

Porsche *Boxster*, 1997–1998

FERDINAND ALEXANDER PORSCHE

BORN 1935
STUTTGART, GERMANY

Ferdinand Alexander "Butzi" Porsche is the grandson of Professor Ferdinand Porsche (1875–1951) who founded **Porsche** AG. He apprenticed as an engineer at **Bosch** in Stuttgart and from 1957 studied at the Hochschule für Gestaltung in Ulm. The following year, he began working under Erwin Komenda (1904–1966) in the Porsche design department, which he later headed from 1961 to 1972. In this position, he designed the bodies of several cars, including the famous *911* (1964). In 1972 he established the independent Porsche Design Studio, now based in Zell am See, Austria. The experience he gained in automotive design proved useful in the creation of "lifestyle" products. His sleek and highly engineered products, which balance function, technology and styling, epitomise German product design. The Porsche Design Studio has produced work for a prestigious international clientele, ranging from the *Cobra* motorcycle for Steyr-Puch (1976) to ergonomic high-tech sunglasses for Bausch & Lomb (1997). During the late 1990s, the studio also produced designs for transit systems for the cities of Singapore, Bangkok, and Vienna and developed a proposal for a computer-aided taxi system (CATS). While most of the products designed by the studio are produced and marketed under the brand name of Porsche Design, the creative origin of other products designed for some of the studio's customers is not always made evident.

Singapore *Airport Express* train for Siemens SGP, 1998

Diver's watch for IWC, 1983

→Motorcycle helmet for Römer, 1976

Telephone for NEC, 1981

TC 91Lz11d1.5100 coffee machine, *TW 91100* cordless kettle & *Tt 91100* toaster for Siemens, 1997

REMINGTON

FOUNDED 1816
ILION GORGE, NEW YORK, USA

C. L. Sholes & C. Glidden, Remington *No.1* typewriter, 1873

In 1816 Eliphalet Remington (1793–1861) fashioned a flintlock rifle that became renowned for its accuracy, and that same year turned his family's forge in Ilion Gorge, New York, into a gun manufacturing business. The venture grew, and in 1828 Remington established a large factory alongside the Erie Canal. Remington and his son Philo subsequently pioneered numerous innovations in arms manufacturing, including a lathe for cutting gun stocks and the reflection method used to straighten the barrels of guns. They also developed the first American practical cast-steel drilled rifle barrel and in 1847 supplied the US Navy with its first breech-loading rifle. As a pioneer of large-scale industrial production, it is not surprising that the company eventually diversified into other areas of manufacture. In 1873 E. Remington & Sons developed the first-ever commercial typewriter, the Remington *No. 1*. This landmark product was designed by Christopher Latham Sholes (1819–1890) and was based on an earlier, cruder model which Scholes had patented in 1868. It was Sholes who first coined the word "typewriter", and his Remington *No. 1* featured the "QWERTY" keyboard layout that is still the standard today. Indeed, many features of this revolutionary machine remained standard for over a century: the cylinder, with its line-spacing and carriage-return mechanism; the escapement mechanism, which moves the carriage along between each letter; the actuation of the typebars by means of key levers and wires; and printing through an inked ribbon. 1878 saw the introduction of the improved Remington *No. 2*, which was the first typewriter to have a shift-key mechanism and was a much better seller than the Remington *No. 1*, which could only print capital letters. Around this time, the company further diversified its product range to include sewing machines. In 1886 E. Remington & Sons sold its typewriter business and became the Remington Arms Company, which supplied a significant number of small arms to the US government during the First and Second World Wars and remains in operation today. Meanwhile, the Remington

Advertisement for Remington sewing machines, c. 1880

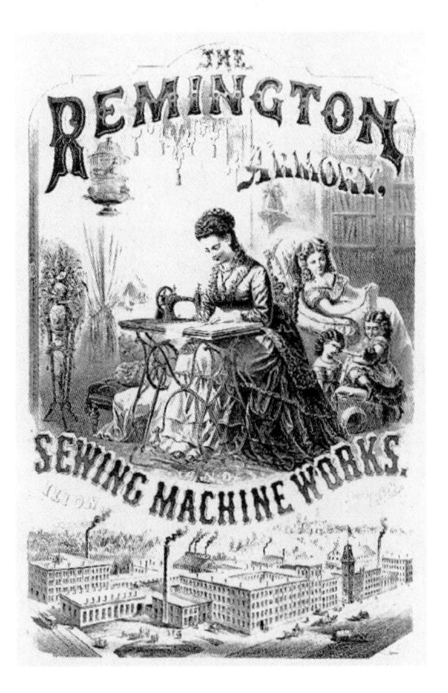

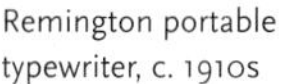
Remington portable typewriter, c. 1910s

typewriter manufacturing company merged with the Rand Kardex Company in 1927 and subsequently became a leading supplier of office equipment. In 1936 it established an electric shaver division and the following year launched its innovative *Close Shaver*. At the 1939 New York World's Fair, Remington introduced the idea of dry shaving to literally millions of Americans, and a year later launched its first dual-headed shaver, the Remington *Dual*. In 1960 Remington brought out the first cordless shaver, the *Lektronic*, employing rechargeable nickel cadmium energy cells, and in 1975 adopted the *Soft Touch* flexible foil cutting system. Four years later Victor Kiam formed Remington Products Inc., stating: "I liked the shaver so much I bought the company". In 1981 the first Remington retail store was opened and over the succeeding decades the company has become a leading developer of shaving, grooming, personal care and travelling appliances.

Renault *R4*, 1961

RENAULT

FOUNDED 1898
BOULOGNE-BILLANCOURT, FRANCE

In 1898 Louis Renault (1877–1944) built his first automobile, the *Voiturette*, in a small workshop on his family's estate in Billancourt. This early design had a direct transmission – an invention that Louis Renault patented. The same year, he founded the Sociéte Renault Frères with his two brothers, Fernand and Marcel. Together they designed pioneering racing cars which won numerous competitions, but when Marcel was tragically killed during a Paris-Madrid race in 1903, the company decided to concentrate its efforts on production vehicles. During the First World War, Renault manufactured a tank that was used as a troop escort vehicle. After the war, Renault's production capacity was increased and during the 1930s the company manufactured a vast array of vehicles, including taxis. Although Louis Renault knew how to make successful cars, he was not so politically astute and continued to produce military equipment during the German occupation of France. Afterwards, he was imprisoned for collaboration and his company was nationalized in 1945. In 1961 the company issued the highly successful Renault *R4* (1961), which in turn became the basis of the landmark Renault 5, introduced in 1972. Designed by Michel Boué (1936–1971), the stylish Renault 5 hatchback was specifically intended for young people and women and created a completely new category of car– the "supermini". By 1975 Renault was producing 1.5 million cars per annum, of which 55 % were exported. Much of this success can be attributed directly to the Renault 5, which became the best-selling French car of all time. A new version of the Renault 5 was launched in

Renault 5 *TL*, 1972

Renault *Espace 2000 TSE*, 1986

Renault *Twingo*, 1992

1982 and in 1986 the company introduced another landmark design, the *Espace*, which similarly created a new market sector – the "people carrier". During the 1990s, the *Twingo* (1992) was one of the first automotive expressions of soft design, while the compact *Scenic* (1996) – voted Car of the Year in 1997 – was the first monospace design to appear in the lower-medium (M1) segment of the market. Today, Renault remains the leading French car manufacturer and continues to be one of the world's most innovative automotive companies.

Design for an industrial washing machine for Electrolux, 1950

SIXTEN SASON

BORN 1912 SKOVDE, SWEDEN
DIED 1967 SWEDEN

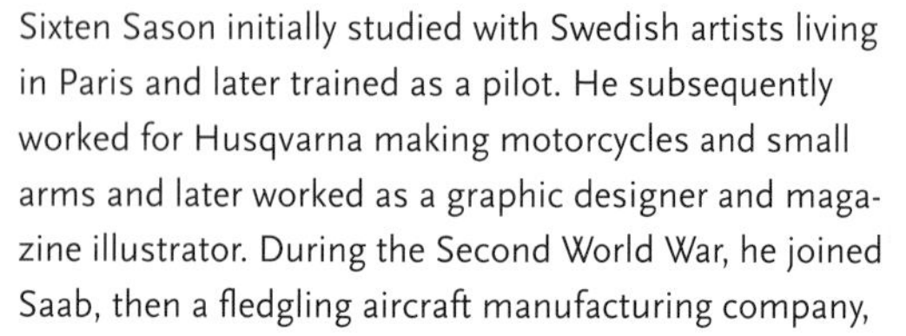

Sixten Sason initially studied with Swedish artists living in Paris and later trained as a pilot. He subsequently worked for Husqvarna making motorcycles and small arms and later worked as a graphic designer and magazine illustrator. During the Second World War, he joined Saab, then a fledgling aircraft manufacturing company, and worked as a technical illustrator. Around 1945 the company decided to diversify, and Saab's chief engineer Gunnar Ljungström and his team began developing its first car, the Saab *92* (1947). Inspired by aeronautical design, Sason transformed this front-wheel-drive vehicle into a streamlined vision with integrated side panels, bonnet and boot – an innovation that was later adopted by Pininfarina and **Raymond Loewy**. The Saab *92* was Sweden's first small car and became the blueprint for later Saab models, such as the *95*, *96* and *99* (all of which were designed by Sason). As one of the first European industrial design consultants, Sason also designed highly influential streamlined products for Electrolux and Hasselblad.

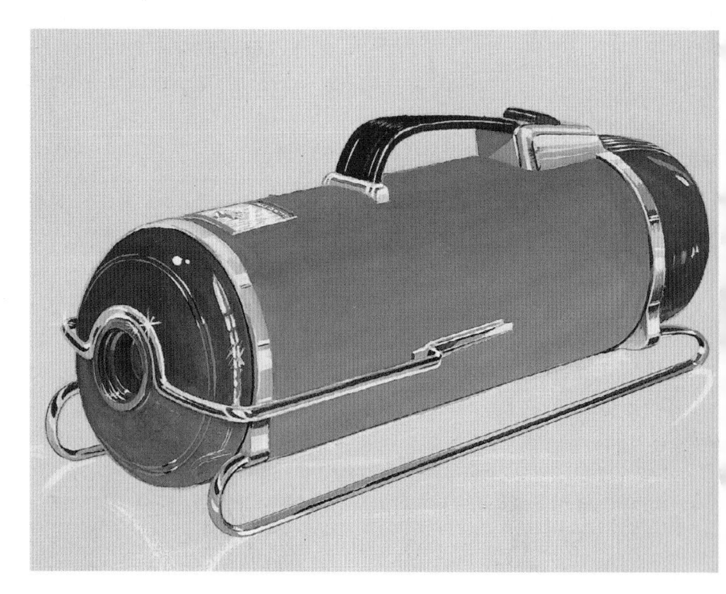

Design for *Model 248* vacuum cleaner for Electrolux, 1943

DOUGLAS SCOTT

BORN 1913 LONDON, GREAT BRITAIN
DIED 1990 LONDON, GREAT BRITAIN

Routemaster bus for London Transport, 1954

Douglas Scott trained as a silversmith at the Central School of Art & Design in London and subsequently worked in **Raymond Loewy**'s London office from 1936 to 1939. While there, he worked on designs for vacuum cleaners and refrigerators commissioned by Electrolux. After the Second World War, he set up his own design office and was responsible for establishing the industrial design course at the Central School of Arts and Crafts in London. In 1949 he co-founded the Scott-Ashford Associates design office with Fred Ashford. He later designed the classic *Routemaster* bus (1954) for London Transport, which is still in use in London today and became a widely recognized and much-loved symbol of the city. The Routemaster (also known as the RM) was the last of a series of buses designed specifically for London and intended to compliment and enhance the look of the capital's streets. Having first entered service in 1956, the *Routemaster* is still considered by many a high-point in public transport design – a triumph of pragmatic engineering that is simple to maintain and operate. With its two-person crew and open back, the *Routemaster* can pick up and set down passengers more quickly than modern driver-only buses. Despite these advantages, the *Routemaster* is unable to comply with recent European health and safety legislation and is slowly being phased out.

Following this successful foray into transport design and engineering, in 1955 Scott established his own independent industrial design consultancy, Douglas Scott Associates. He subsequently designed products for companies such as Marconi, Ideal Standard and ITT. Scott believed that there was no place for personal aesthetics in industrial design. His telephone coin box (c. 1960) for the GPO (General Post Office), which combines American styling with the traditional restraint of British design, is highly representative of his work.

Pay-on-answer coin box for the General Post Office, c. 1960

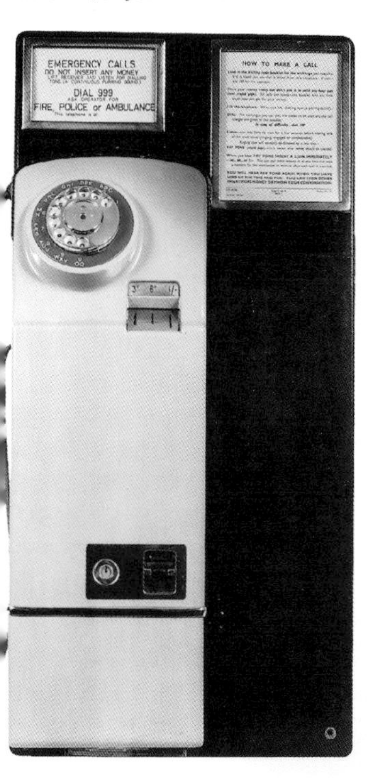

HANS ERICH SLANY

BORN 1926
ESSLINGEN, GERMANY

BR Trike for Kärcher, 1990s

Hans Erich Slany studied engineering in Eger (Hungary) and in Esslingen. Between 1948 and 1955 he developed products for Ritter Aluminium, before establishing his own design studio, Slany Design, in 1956. Around this time he collaborated with Heinrich Löffelhardt (1901–1979), co-designing the *Ikonette* compact camera for Zeiss-Ikon (1956). In 1959 Slany was one of the founders of the Verband Deutscher Industrie-designer (German Association of Industrial Designers), and that same year designed his first electric drill for **Bosch**. Utilizing ergonomics data and shock-proof and heat-resistant thermoplastics, Slany went on to design numerous other power tools for Bosch, including an electric screwdriver (c. 1960), the *Combi-E* range of DIY tools (1966) and the first electro-pneumatic rotary hammer weighing less than 2.5 kg (1981). Since 1962 Slany has designed over 200 high-pressure cleaning products for Alfred Kärcher GmbH and has also designed a wide variety of other products, ranging from medical equipment to electronic rocketry components. He was made an honorary professor at the Berlin Hochschule der Künste in 1985. Over its 40-year history, Slany's office – which in 1997 renamed itself Teams Design – has won more than 900 national and international design awards, more than any other design consultancy in the world. Slany believes that "individual product personality" is essential for success in today's increasingly homogenous marketplace.

GBH 24 VRE drill for Bosch, c. 1992

TR-55 transistor radio, 1955 – Japan's first transistor radio

SONY

FOUNDED 1946
TOKYO, JAPAN

In the period immediately following the Second World War, Japan saw a huge surge in demand for radios, fuelled by a populace desperate for news from around the world. In September 1945 the engineer Masaru Ibuka responded to this opportunity by opening a small electrical repair shop in Tokyo. Called the Tokyo Tsushin Kenkyujo (Tokyo Telecommunications Research Institute), the business repaired war-damaged radios and shortwave units and made its own shortwave adapters and converters that could turn short-wave radios into all-wave receivers. Ibuka was joined by his friend, the physicist Akio Morita (1921–1999), and together they founded the Tokyo Tsushin Kogyo Kabushiki Kaisha (Tokyo Telecommunications Engineering Corporation) in May 1946. Although the company's best-selling product was initially an electrically heated cushion, in 1950 – by now in larger premises – it introduced the first Japanese magnetic tape, the *Soni-Tape*, and also began marketing Japan's first reel-to-reel tape recorder, the *G-Type*. Later, the company developed the *H-type*, a less bulky and easier-to-use tape recorder in a case, which was more suitable for home use and was especially popular for educational purposes. In 1952 it developed a stereophonic audio system for the first Japanese stereo broadcast by NHK. That same year, Ibuka visited the United States and while there discovered that Western Electric was planning to release rights to its transistor patent to companies prepared to pay royalties – the transistor having been previously developed in 1948 by scientists at Bell Laboratories. Ibuka's company managed to obtain a licence to manufacture the transistors in 1954 and the following year introduced the first Japanese transistor radio, the *TR-55*. This was followed by the

TV8–301 television, launched in 1959 – the world first's portable television

SONY
WALKMAN
MIC
STOP
EJECT
LEFT
RIGHT
VOLUME
SONY

world's first pocket-sized transistor radio, the *TR-63* (1957). In 1958, aware of the need to appeal not just to Japanese consumers but to a global audience, the company changed its name to Sony Corporation, which sounded more Western. Around this time Ibuka noted: "The days of radio are over. The future lies in television", and in 1959 the company launched the first-ever transistor television, the *TV8–301*. Two years later, Sony entered into a contract with Paramount Pictures that involved the studio providing "technical assistance in the production of a chromatron tube and colour television utilizing it". This agreement led to the development of the small but revolutionary *Trinitron* colour televisions, which were first introduced in 1968. They included the *TV5–303*, the smallest and lightest micro-television in the world, which created a boom in America for micro-TVs. Sony's pioneering miniaturization of electronic technology continued with the introduction of the first *Walkman* personal stereo in 1979. With excellent and unwavering sound reproduction, the light and highly portable *Walkman* was an instant success and initially Sony could not keep up with consumer demand. Unlike conventional stereos with large speakers, the *Walkman* needed only a small amount of battery power because the sound was directed straight into the listener's ears. Sony has subsequently produced many versions of this landmark design, including models suitable for sporting activities. It went on to launch a flat-screened portable television, the *Watchman*, in 1982, and a portable compact disc player, the *Discman*, in 1990. More recently, Sony has been developing various digital technology products, including its miniature *Memory Stick*, which provides compact portable storage for digital data, and the sound-responsive *AIBO* robotic dog, which has infra-red eyes and can be trained to perform tricks. Sony predicts: "The Eighties was the age of the PC and the Nineties was the age of the Internet, the 2000s will be the age of the robot."

Voyager Watchman, 1982

←*Walkman*, first launched in 1979

AIBO robotic dog, 1999

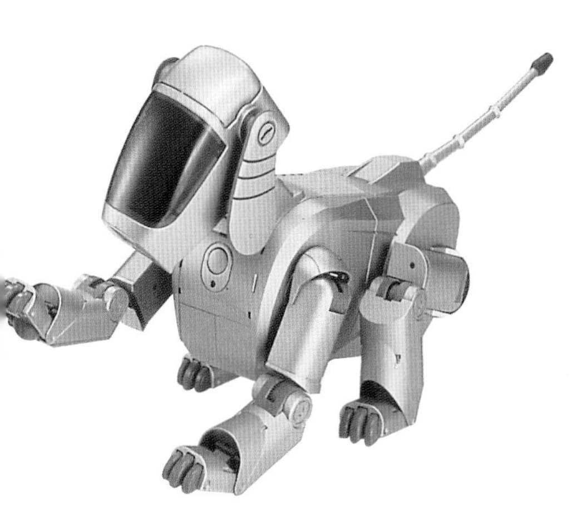

Detail of *Model 1865* rifle

SPRINGFIELD ARMORY

FOUNDED 1794
SPRINGFIELD, MASSACHUSETTS, USA

In 1777, during the American Revolution, an arsenal was built in Springfield to produce cartridges and gun carriages. Later, the new Federal Government decided to begin manufacturing its own muskets so that it would no longer have to rely on imported arms. To this end, it established a new Armory in Springfield in 1794 and a year later the production of muskets began, with 40 workers producing 245 flintlock muskets per month. In 1819 Thomas Blanchard invented a special lathe which enabled rifle stocks to be mass-produced to a standardized design. In the period running up to the outbreak of the American Civil War, the Armory became a beacon of the Industrial Revolution. During the 1840s, the flintlock was replaced by a percussion cap and hammer, an innovation which led to better weather performance and faster firing. The development of rifling – cut spiral grooves in a barrel – to spin projectiles also ensured greater range and accuracy. The Union Army's subsequent victory at Gettysburg in 1863, and indeed in the Civil War as a whole, can to some extent be attributed to the *Springfield* rifle (1861) and the unprecedented scale of its mass production: between 1863 and 1864, the Springfield Armory boosted its productivity to a phenomenal 25,000 weapons per month using rationalized manufacturing techniques such as the division of labour, mechanization where possible and standardization. In testimony to the influence of the Springfield Armory and other pioneering small arms manufacturers such as **Eli Whitney**, mass production in America was initially referred to as "armory practice" when it was first employed by automobile manufacturers in Detroit. The Springfield Armory also produced the well-known *M1* rifle (1926), also called the *Garand* Rifle after its inventor John Garand, of which 4.5 million were made. Today, in recognition of its contribution to both American history and the evolution of industrial manufacture, the Springfield Armory has been designated a National Historic Site.

Springfield rifle with telescopic sight, 1903

WALTER DORWIN TEAGUE

BORN 1883 DECATUR, INDIANA, USA
DIED 1960 FLEMINGTON, NEW JERSEY, USA

Cash register for the National Cash Register Company, c. 1937

Walter Dorwin Teague was one of the great pioneers of industrial design consulting in America. Having moved to New York in 1903, he subsequently attended evening classes at the renowned Art Students League. He later worked as an illustrator for a mail-order catalogue and also for the Hampton Advertising Agency. In 1912 he established his own studio in New York, working as a typographer and graphic designer, and in the mid-1920s also began designing packaging. In 1926 he founded Walter Dorwin Teague Associates, which was one of the earliest industrial design consultancies. Teague's first client was Eastman **Kodak**, for which he undertook a comprehensive design programme that included design research and product development. His first camera, the *Vanity Kodak* (1928), was designed specifically for the female market and was produced in various colours with matching silk-lined cases. Teague later received widespread acclaim for his *Baby Brownie* camera (1933), which was one of the first consumer prod-

Bantam Special camera for Kodak Company, 1933–1936

→Automatic accounting machine for the National Cash Register Company, c. 1950

↘*Edgemount Combination Barometer* for Taylor Instrument Companies, c. 1950

Label design for soft-drink bottles for Nehi Corporation, c. 1950

ucts made of plastic. His best-known camera, however, was the distinctively styled *Bantam Special* (1936), which was more compact and user-friendly than earlier Kodak models. In 1930 Teague designed the streamlined body of the Marmon *Model 16* car, which was among the most aerodynamically efficient automobiles of its time. He also designed other streamlined products, including glassware for Corning, kitchenware for Pyrex, pens and lighters for Scripto, lamps for Polaroid, mimeographs for A. B. Dick, radios for Sparton and the *Centennial* piano for Steinway. As well as consumer prod-

Mock-up of an interior for the Boeing 707, 1956

ucts, Teague also designed a plastic truck body for UPS, supermarkets for Colonial Stores, interiors for the **Boeing** 707 airliner, United States pavilions at various international trade fairs, exhibition interiors for **Ford**, service stations for Texaco and various exhibits at the 1939 New York World's Fair, including a gigantic cash register, which recorded visitor numbers and was based on his earlier design for the National Cash Register Company. He also wrote *Design This Day – The Technique of Order in The Machine Age* (1940), which celebrated the potential of machines and the "new and thrilling style" of the Modern era. Now headquartered in Redmond, Washington, with a regional office in California, Walter Dorwin Teague Associates Inc. employs a staff of 200 professionals and specializes in transportation design, systems engineering, human factors and corporate identity.

Jack St. Clair Kilby holding up an early integrated circuit, c. 1959

TEXAS INSTRUMENTS

FOUNDED 1951
DALLAS, TEXAS, USA

Texas Instruments originated from a company, Geophysical Service, established in 1930 by Clarence "Doc" Karcher and Eugene McDermott. Geophysical Service specialized in the reflection seismograph method of exploration and, later, submarine-detection equipment and airborne radar systems. In 1951 it became a wholly-owned subsidiary of the newly-founded Texas Instruments. In 1952 Texas Instruments acquired a licence from the Western Electric Company to manufacture transistors, and in so doing entered the semiconductor industry. In 1958 Texas Instruments' in-house inventor, Jack St. Clair Kilby, demonstrated the first-ever integrated circuit (IC) and three years later the company delivered the first integrated-circuit computer to the US Air Force. 1964 saw the introduction of the first consumer product to incorporate an integrated circuit – a hearing aid – and the same year the company began mass-producing the first plastic-packaged ICs. Over the next decade, Texas Instruments was at the forefront of electronic technology, developing the first electronic hand-held calculator (1967) and the first laser-guided missile system (1969). In 1971 the firm pioneered the single-chip

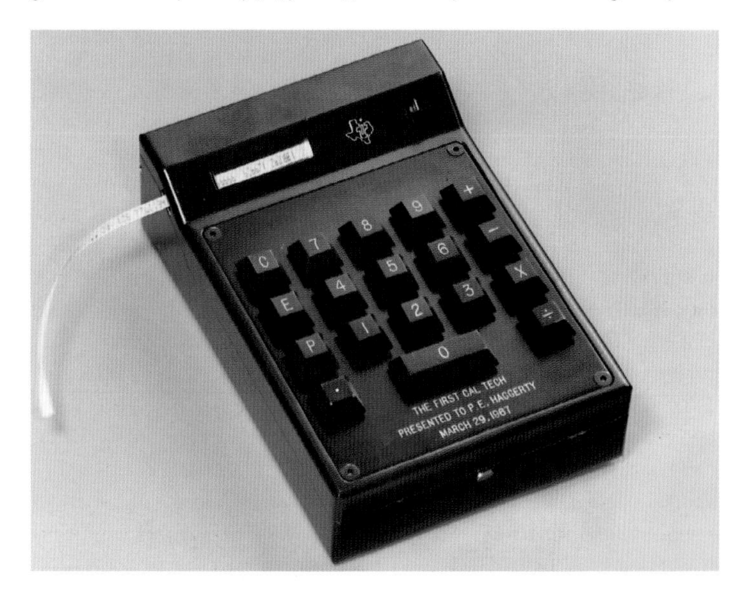

The first electronic hand-held calculator, developed by Jack St. Clair Kilby, Jerry D. Merryman and James Van Tassel, 1967

Speak & Spell
learning aid, 1978

microprocessor and the single-chip microcomputer, and a year later entered the consumer market with its *Datamath* hand-held calculator. In 1974 it launched the *TMS1000* one-chip microcomputer and the following year introduced three-dimensional data-processing technology. The first single-chip speech synthesiser was unveiled in 1978, as was the first consumer product to incorporate low-cost speech synthesis technology, the *Speak & Spell* learning aid. In 1987 the company pioneered the first single-chip 32-bit artificial-intelligence microprocessor, and four years later became the first American manufacturer of semiconductors to establish a research and development facility in Japan. Over the years, Texas Instruments has won numerous awards for manufacturing excellence and its many inventions, which have driven completely new industrial design typologies.

Michael Thonet, *Boppard Chair I*, 1836–1840

MICHAEL THONET

BORN 1796 BOPPARD, GERMANY
DIED 1871 VIENNA, AUSTRIA

Michael Thonet established a furniture workshop in Boppard am Rhein in 1819. From around 1830, he began experimenting with laminated wood and produced a number of innovative chairs in the Biedermeier style. These designs prompted the Austrian chancellor, Prince Metternich, to invite Thonet to Vienna, where he was granted a patent for his new process for bending wood laminates in 1842. Having secured the necessary financial backing, Thonet moved to Vienna and in 1849 established a furniture workshop in Gumpendorf, a suburb of Vienna, with his sons, Franz, Michael, August and Joseph. For the next two years, the Thonet family concentrated on developing techniques for mass-producing furniture, including the steam bending of solid wood. In 1851 Thonet exhibited its new furniture designs at the Great Exhibition in London and won a gold medal. By 1853 Gebrüder Thonet, as the company was now known, had moved into larger premises and was fully

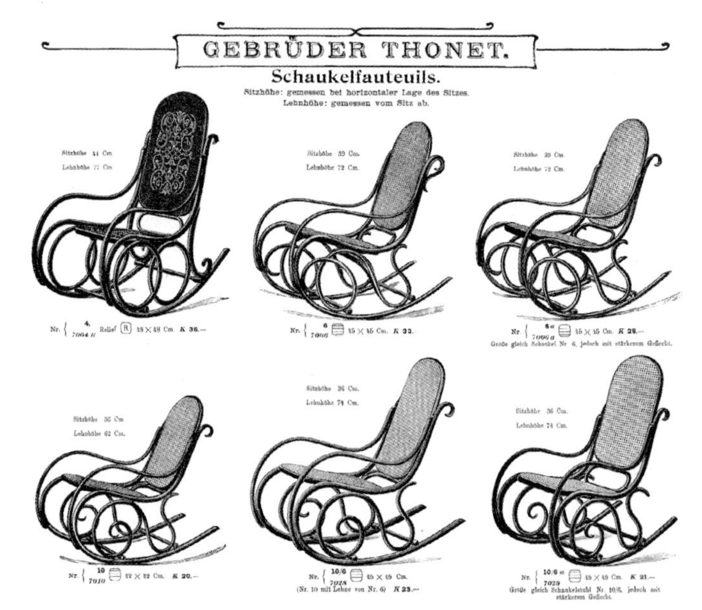

GEBRÜDER THONET.

Schaukelfauteuils.

Sitzhöhe: gemessen bei horizontaler Lage des Sitzes.
Lehnhöhe: gemessen vom Sitz ab.

Sitzhöhe 41 Cm.
Lehnhöhe 77 Cm.
Nr. { 4 / 7004 R Relief (R) 48 × 48 Cm. K 36.—

Sitzhöhe 39 Cm.
Lehnhöhe 72 Cm.
Nr. { 6 / 7006 45 × 45 Cm. K 33.

Sitzhöhe 39 Cm.
Lehnhöhe 72 Cm.
Nr. { 6 a / 7006 a 45 × 45 Cm. K 28.—
Größe gleich Schaukel Nr. 6, jedoch mit stärkerem Geflecht.

Sitzhöhe 36 Cm.
Lehnhöhe 62 Cm.
Nr. { 10 / 7010 42 × 42 Cm. K 20.—

Sitzhöhe 36 Cm.
Lehnhöhe 74 Cm.
Nr. { 10/6 / 7028 45 × 45 Cm.
(Nr. 10 mit Lehne von Nr. 6) K 23.—

Sitzhöhe 36 Cm.
Lehnhöhe 74 Cm.
Nr. { 10/6 a / 7029 45 × 45 Cm. K 21.—
Größe gleich Schaukelstuhl Nr. 10/6, jedoch mit stärkerem Geflecht.

Page from Thonet catalogue, 1904

Model No. 4 side chair, c. 1850

mass-producing chairs distinguished by a reduction of elements and the elimination of extraneous ornament. In 1856 the company opened its first factory in Koritschan, Moravia. Gebrüder Thonet's remarkable success was due to its adherence to mechanized methods of production, which allowed it to sell its products at very competitive prices. In 1860, for example, the firm's best-known model, the *No. 14* chair, cost less than a bottle of wine, and by 1891 a staggering 7,300,000 of these ubiquitous café chairs had been sold. In the early 1900s several leading Viennese architects, including Josef Hoffmann (1870–1956), began designing furniture for Gebrüder Thonet in the Secessionist style. In 1929 a French subsidiary was established, Thonet Frères, which produced progressive tubular metal furniture designed by Marcel Breuer (1902–1981) and others. The production of this furniture was later moved to Frankenberg, Germany. Thonet continues to operate and manufactures re-issues of its earlier seating as well as contemporary designs.

TUPPERWARE

FOUNDED 1938
LEOMINSTER, MASSACHUSETTS, USA

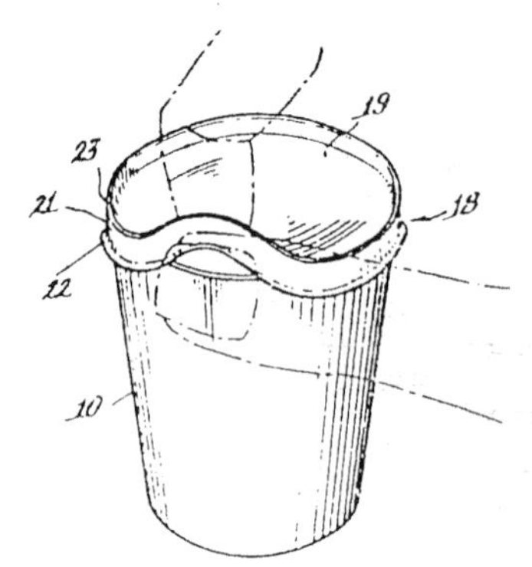

Drawing showing the Tupper Seal, patented in 1947

Earl Tupper (1907–1983) gained his love of invention from his father, who used to construct labour-saving devices on the family farm. In 1936 Tupper met Bernard Doyle, who had developed a plastic known as Viscoloid, and he subsequently took a job in the plastics manufacturing division of DuPont. Although he only stayed there a year, Tupper later recalled that it was where "my education really began". In 1938 Tupper established his own plastics manufacturing company in Leominster, Massachusetts. The company initially undertook subcontracted work for DuPont, but subsequently won its own contracts for military products such as gas masks and signalling lamps. During the immediate post-war years, Tupper turned his attention to the design of plastic consumer products, such as sandwich picks, bathroom tumblers and cigarette cases, which firms could give away to their customers as promotional sales tools. Plastics in the 1940s still had numerous unpleasant properties, ranging from brittleness and odour to greasiness. To overcome these problems,

Single-impression mould used in the injection-moulding of the *Mix-n-store* container, 1965

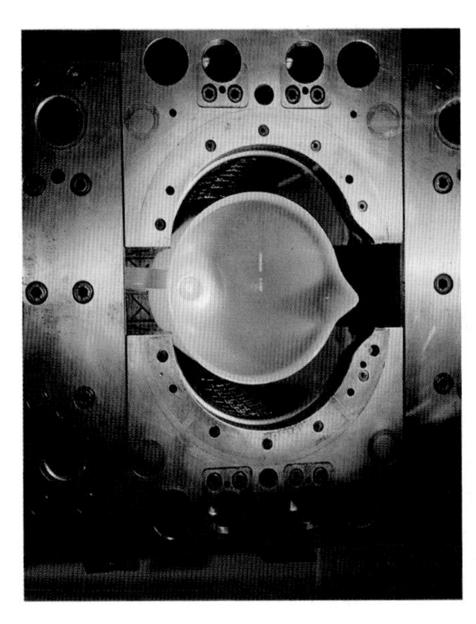

Oyster, multi-purpose storage containers, 1990s

Tupper developed a technique for purifying black polyethylene slag (a by-product of the oil industry) into a strong, flexible, non-porous material that was translucent and non-greasy. Around the same time, he also patented his famous airtight and watertight Tupper seal, the design of which was based on the lids used for paint containers. Tupper combined these two innovations to create a range of airtight plastic containers, but found that the product line "fell flat on its face" in retail stores. He realised that his sealing system had to be demonstrated before consumers would buy into it. In the late 1940s Stanley was successfully selling its household products through local distributors, who arranged demonstration home parties, and in 1948 Tupper met several of these distributors. Amongst them was Brownie Wise, who expanded this novel retailing concept for Tupper so that his food storage containers were only available through such demonstrations. From 1951 to 1958 Wise was vice-president of the company, which was then called Tupperware Home Parties. As a skilful saleswoman and motivator of people, Wise understood the importance of her party hostesses, declaring: "If we build the people, they'll build the business." With her large sales force, she transformed the company's fortunes and Tupperware parties became as famous as Tupperware products. Today Tupperware produces a vast range of high-quality plastic housewares, from kitchen equipment and multi-purpose domestic storage containers to organizing products. A Tupperware demonstration party is held on average every two seconds somewhere in the world.

←Advertisement showing a Tupperware party scene, 1960s

HAROLD VAN DOREN

BORN 1895 CHICAGO, ILLINOIS, USA
DIED 1957 PHILADELPHIA, PENNSYLVANIA, USA

Harold van Doren was a pioneer of industrial design consulting and streamlining who designed products with a clean, contemporary aesthetic. He first studied languages before moving to Paris, where he worked at the Louvre. After returning to the United States, he worked as an assistant to the director of the Minneapolis Institute of Arts, but resigned from this position when he was given the chance to work in the fledgling field of industrial design. One of his first commissions came from the Toledo Scale Company, who asked him to design some commercial weighing scales. Van Doren's lightweight and innovative solution was one of the first products to incorporate a large-scale plastics moulding. Having based his practice in Philadelphia, he went on to design many streamlined products for Maytag, Goodyear, Ergy, Swartzbaugh and DeVilbiss. Together with John Gordon Rideout, he designed the widely acclaimed green skyscraper-shaped plastic

First range manufactured by Philco, 1950

Magnalite tea kettle for Wagner Ware, 1940 (co-designed with J. G. Rideout)

ʼadio for Air-King (1930–1931) and a child's scooter (1936). In 1940 van Ɔoren wrote *Industrial Design: A Practical Guide to Product Design and Development*, and nine years later published an article in *Design* magazine entitled "Streamlining Fad or Function?". Here he argued that the curved ines common to refrigerators at the time were the result of a manufacturing process involving a metal press known as a "bulldozer", and that the formal anguage of the product was thus "imposed on the designer by the necessity of obtaining low cost through high-speed production". Van Doren was par-ticularly adept at re-styling products without altering their existing layout, as illustrated by his cooking range (1950) for the **Philco** Corporation. This streamlined design was a re-modelled version of an earlier range produced by Electromaster Inc. that "even experienced men in the industry believed o be entirely new". For Philco, van Doren also designed a streamlined re-frigerator (c. 1950) whose door was emblazoned with the words "Advanced Design" – an early example of the use of design as a marketing tool.

Ferdinand Porsche, original Volkswagen *Beetle*, 1938 – first designed in 1934

VOLKSWAGEN

FOUNDED 1938
WOLFSBURG, GERMANY

In January 1934 the automotive designer Professor Ferdinand **Porsche** submitted a design proposal to the new German Reich government for a revolutionary "car for the masses", called the *Volkswagen* (People's Car). Soon afterwards he signed a contract with the Reichsverband der Automobilindustrie (Automobile Industry Association), which provided him with a development budget on the understanding that a prototype of the car would be completed within ten months. In his pursuit of technological excellence, Porsche introduced innovations into the chassis, engine and transmission, regardless of cost; indeed, despite the availability of cheaper alternatives, he selected the more expensive air-cooled, horizontally-opposed engine for the car because of its full-throttle endurance. The first *Volkswagen* prototype was unveiled in October 1935, and by the following spring successive prototypes were undergoing extensive testing. The earliest cars, with their modern all-steel bodywork, were later improved with the addition of bumpers and running boards. Erwin Komenda (1904–1966), the acclaimed aerodynamicist, designed a distinctive streamlined body for the car, which, although compact, could easily accommodate five passengers. In 1938, in advance of the full-scale mass-production of the *Volkswagen*, the Gesell-

Volkswagen-Werk GmbH brochure, c. 1937, designed by Thomas Abeking

Volkswagen *Beetle*, model of the late 1970s (first re-designed in 1945)

Volkswagen *New Beetle*, 1998

schaft zur Vorbereitung des Volkswagens (Company for the Preparation of the People's Car) was established, later renamed Volkswagen-Werk GmbH. When a prototype of the *People's Car* was shown to the press later the same year, it was immediately dubbed the "Beetle" by the *New York Times*. The German government, on the other hand, preferred the propagandist "Kraft durch Freude" (Strength through Joy) name for the new car, which was initially intended to be sold through a savings stamp scheme. Work began on the construction of the world's largest car plant, where the cars were to be manufactured, and a town for workers was also built (the Town of Strength

Volkswagen *Transporter* from the 1970s (first designed in 1950)

Giorgetto Giugiaro, Volkswagen *Golf Mk 1*, 1974

through Joy, today Wolfsburg). In 1940 the still incomplete factory started producing armaments, and *Volkswagen* production took a back seat while Porsche converted the car into two military jeep-style vehicles called the *Kübelwagen* and the *Schwimmwagen*. A year later, serial production of the *Volkswagen* was initiated with the manufacture of 41 cars, which were mainly used by the Nazi Party for propaganda purposes. After the war, *Volkswagen* production accelerated considerably, with the British Military Authorities placing an order for 20,000 vehicles in September 1946. Some 100,000 *Volkswagens* had been manufactured by 1950, the year in which another "classic" design was launched – the VW *Transporter*. This famous camper van was developed from a rough sketch by a Dutch VW importer who was inspired by the "flat vehicles" used in factories. Functioning as either a mini-

Volkswagen *Passat V6 TDI*, 2000

bus or a commercial vehicle, the *Transporter* was hugely successful, and by 1951 Volkswagen was producing some 12,000 units per annum. The *Karmann Ghia Coupé* was launched in 1955 and two years later a sophisticated cabriolet version was unveiled at the Frankfurt International Motor Show. By 1962 Volkswagen was producing 1,000,000 cars per annum. In 1964 the company built a huge research and development complex in Wolfsburg, which included a state-of-the-art wind tunnel opened in 1965. With this facility, VW was able to increasingly use technological advances as the basis for its future designs. In 1972 the *Beetle* became the world's most produced car. Volkswagen launched the *Passat* in 1973, followed in 1974 by the *Golf* designed by Giorgetto Giugiaro (b. 1938). Both cars served as blueprints for subsequent generations of Volkswagen models. The *Golf* – the intended successor of the *Beetle* – proved one of Volkswagen's greatest successes: one million of these "classless quality cars" were manufactured within the first 31 months and in 1977 the *Golf GLS* received a "Gute Form" award sponsored by the Federal Ministry of Economic Affairs. By 1988 an impressive ten million *Golfs* had been produced. Volkswagen's highly successful *New Beetle* was launched in 1998 and immediately received numerous awards and accolades, including being named "The Best Design of 1998" by *Time* Magazine. Based on the new *Golf* chassis, this superb design – like all other Volkswagens – "expresses the German engineering passion for designing and building cars". The Volkswagen Group now owns **Audi**, Bugatti, Lamborghini, Seat and Skoda and is one of the largest car producers in the world.

Sir Joshua Reynolds, *Josiah Wedgwood*, 1783

JOSIAH WEDGWOOD

BAPTIZED 1730 BURSLEM, STAFFORDSHIRE, ENGLAND
DIED 1795 ETRURIA, STAFFORDSHIRE, ENGLAND

The youngest of twelve children, Josiah Wedgwood came from a family whose members had been potters since the 17th century. After the death of his father in 1739, he worked for the family pottery, which had been inherited by his eldest brother, Thomas. After a five-year apprenticeship there, in 1754 Wedgwood entered into partnership with an established potter, Thomas Whieldon. Around this time, he began carefully recording his experiments, which included his highly successful formula for green glaze. In 1759 he established his own business at Ivy House Works, where he produced "a species of earthenware for the table, quite new in appearance, covered with a rich and brilliant glaze". A tea and coffee service of this warm cream-coloured earthenware was ordered by Queen Charlotte and in 1765 Wedgwood was given permission to re-christen it *Queen's ware* and to call himself "Potter to Her Majesty". Well finished and clean in appearance, *Queen's ware* was extremely versatile, in that it could be left plain, incised with decoration, hand-painted or transfer printed. Its durability and its serviceable

Wedgwood's Etruria factory opened on 13 June 1769

forms also made it highly practical, and it became the standard domestic pottery selling successfully around the world. The success of *Queen's ware* was not only the result of the increasing demand for pottery fuelled by the growing popularity of tea-drinking, but also due to its simple forms, which were utterly in tune with the emerging Neo-Classical style. To meet the demand for his high quality wares, in 1764 Wedgwood moved his business to the larger Brick House Works in Burslem. Here he pioneered the logical division of labour, employed rationalized production methods and used innovative marketing techniques, while the forms of his cream-ware became less and less ornate so as to be better suited to mass-production. Wedgwood also discovered that cream-coloured earthenware could be successfully coloured with oxides to imitate stones, most notably agate, porphyry, granite and Blue John. In 1768 he entered into partnership with Thomas Bentley, a merchant from Liverpool, to manufacture ornamental wares using this type of coloured earthenware in the Neo-Classical style. A year later, they established the famous canal-side Etruria factory to produce these

Pair of white *Jasper ware* vases with green and lilac decoration, 1862 (large) and 1871 (small)

Keith Murray, coffee service, c. 1934

new moulded wares. The pieces were so popular that the company's London agent was, according to Wedgwood, "mad as a March hare for Etruscan vases". After literally thousands of experiments, in 1774 Josiah Wedgwood finally perfected his cameo-like *Jasper ware*. Many leading artists, including George Stubbs (1724–1806) and John Flaxman (1755–1826), designed relief decorations for *Jasper ware*, which was unashamedly Neo-Classical in style. Wedgwood explored every kind of shape and function possible in ceramics. He also investigated new industrial methods of production and in 1782 his Etruria factory became the first to install a steam-powered engine for the mass production of his moulded wares. During the 19th century, the Wedgwood factory continued manufacturing both domestic and ornamental wares. In the 1930s and 1940s, the factory began producing Modern wares designed by Keith Murray (1892–1981) and Eric Ravilious (1903–1942), among others. Alongside its more traditional wares, Wedgwood today continues to produce designs that combine simple forms with modern decoration.

ELI WHITNEY

BORN 1765 WESTBORO, MASSACHUSETTS, USA
DIED 1825 NEW HAVEN, CONNECTICUT, USA

One of the great pioneers of the Industrial Revolution in America, Eli Whitney studied science and technology at Yale College (now Yale University). After graduating in 1792, he accepted a teaching post in South Carolina, but was shipwrecked on his way there and ended up in New York, where he met Phineas Miller and Catherine Greene. Travelling south with them, he stayed at Greene's plantation, Mulberry Grove, where he learned of the need for a new machine that would help remove seeds from the short-fibred cotton that grew in the South, and which unlike long-fibred cotton could not be easily cleaned. Within six months, Whitney had constructed a working model of his revolutionary cotton gin. "This machine may be turned by water or with a horse, with the greatest of ease", he declared, "and one man and a horse will do more than fifty men with the old machines". The device pulled the cotton through hundreds of closely set teeth mounted on a revolving cylinder that combed the seeds out. The fibre was then passed through slots in an iron breastwork that were narrow enough not to let seeds through. Whitney's cotton gin had an enormous impact on the South's economy, which became almost completely cotton-based. This simple yet ingenious design was patented in 1794. Whitney subsequently

Engraving of Eli Whitney's cotton gin, designed in 1793

Advertisement showing Eli Whitney's fire arms, mid-19th century

went into business with Miller to begin mass-producing the cotton gin using purpose-built machinery. The patent rights to the device, however, were widely infringed by planters who made their own copies, which resulted in Miller & Whitney going out of business in 1797. Having learned from this experience, Whitney decided to turn his attention to the manufacture of small arms, as the US Government, fearing the outbreak of war with France, was attempting to solicit 40,000 muskets from private contractors. Muskets at this stage were still made almost entirely by hand and individually, so that each one differed slightly from the next. Whitney resolved to produce small arms using specially-designed machinery allowing for the mass-manufacture of interchangeable parts. This novel and far-reaching concept of standardization was completely revolutionary and had enormous implications for later industrially produced designs. Standardized interchangeable parts enabled faster assembly and much easier replacement of broken elements. Funded by the US government, Whitney established an armoury in New Haven to produce his famous muskets. There he almost single-handedly pioneered what became known as the "American System" of mass-production, which by the late 1900s was being used for the large-scale manufacture of numerous products, from sewing machines and clocks to automobiles.

WILBUR & ORVILLE WRIGHT

BORN 1867 NEAR MILLVILLE, INDIANA, USA	BORN 1871 DAYTON, OHIO, USA
DIED 1912 DAYTON, OHIO, USA	DIED 1948 DAYTON, OHIO, USA

Wilbur (top) & Orville Wright

Having grown up in a home where "there was always much encouragement to children to pursue intellectual interests; to investigate whatever aroused curiosity", the Wright brothers initially established a printing shop where they designed and built their own presses. In 1892 they opened a shop for the sale and repair of bicycles and four years later began constructing bicycles to their own designs. They used the profits from these two ventures to fund aeronautical experiments, whereby the experience gained from the design and construction of the presses and bicycles, and their subsequent understanding of the inherent characteristics of materials such as tubular metal, wood and wire, was of enormous benefit to their research. After corresponding with the French civil engineer and leading authority on aviation, Octave Chanute, in 1900, the Wright brothers realized that a successful flying machine would need wings to generate lift, a propulsion system and a control system. Of these, control posed the greatest problem. Having determined that the best means of control would be through the precise manipulation of the centre of pressure on the wings, they devised a method of mechanically inducing a helical twist across the wings in either direction. This provided lift on one side and decreased lift on the other so that the pilot could raise and lower either wing tip at will. This concept of "wing-warping" was used to control the Wright's first glider of 1900, which was mainly flown as a kite without a pilot on board. Having constructed and flown another glider in 1901, they subsequently built a small wind tunnel in which they tested the performance of over 100 wing designs. Their researches led to the very successful 1902 glider with an improved control system comprising a forward monoplane elevator and a moveable rudder linked to the wing-warping system. First tested in late 1902 at Kill Devil Hills, near the village of Kitty Hawk, North Carolina, the 1902 Wright glider demonstrated that the brothers had overcome the key problems of heavier-than-air flight. So important was this aircraft that the Wright's later basic patent featured a glider

Drawing of Wright brothers' basic patent, 1904

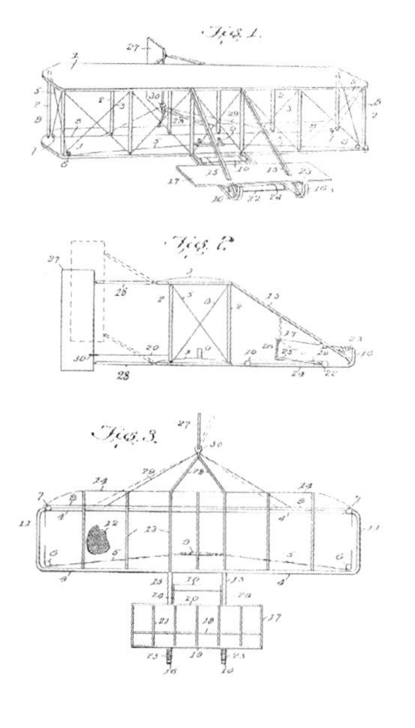

The Wright Brothers' first flight on 17 December 1903

design with the control system of the 1902 model, rather than a powered aircraft. Having made considerable technical progress with their gliders, the brothers moved on to powered machines and built their 1903 *Flyer*. This aircraft was powered by a 12.5 horsepower four-cylinder engine of the Wright's own design, which was linked through a chain-drive transmission to twin counter-rotating pusher propellers. On 17 December 1903 at Kill Devil Hills, with Orville as pilot, this aircraft achieved the first ever controlled, powered flight, flying 120 feet in 12 seconds. Over the next couple of years the Wrights refined their designs so that by 1905 their *Flyer* could remain in the air for 39 minutes. Three years later, the Wrights were contracted by the US government to supply a military aeroplane capable of flying for more than one hour at an average speed of 40 mph. In 1909 the US Army conducted trials of the resulting *Model A*, which was the world's first military aeroplane. That same year, the Wright Company was incorporated, an aeroplane factory was established in Dayton and a flying school was founded at Huffman Prairie. In tribute to the Wright brothers' historic achievements, the label beside their famous 1903 *Flyer* on display in the Smithsonian Institute reads: "By original scientific research, the Wright brothers discovered the principles of human flight. As inventors, builders and flyers, they further developed the aeroplane, taught man to fly, and opened the era of aviation."

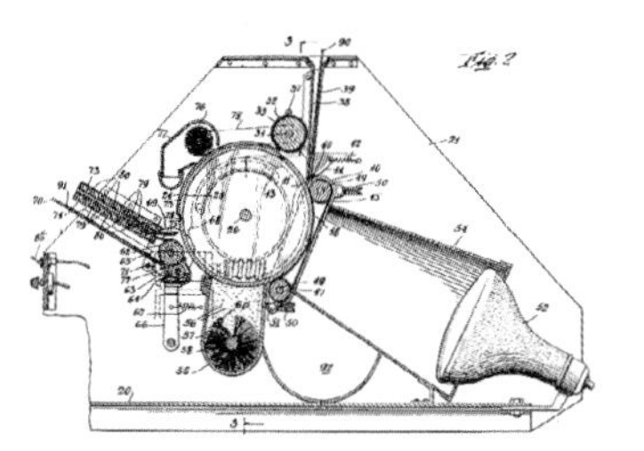

British patent drawing, 1952 – the American patent was originally filed in 1940

XEROX

FOUNDED 1906
ROCHESTER, NEW YORK, USA

Chester Carlson (1906–1968) briefly worked for the Bell Telephone Company, prior to joining the New York-based electronics firm P. R. Mallory Company. While working there as a patent attorney, Carlson became increasingly frustrated by the difficulties he experienced in obtaining copies of patent drawings and specifications. From 1934, he began exploring techniques for copying both text and drawings using electrostatic methods rather than photographic or chemical methods, which were already being researched by a number of large corporations. Having set up a makeshift laboratory in Queens, New York, he developed a prototype photocopier that produced the first xerographic image in 1939. Carlson then spent several years trying to sell his invention to various large corporations, including **General Electric** and **IBM**, but was met with what he described as "an enthusiastic lack of interest". The reasons for this were manifold – Carlson's prototype copier was both bulky and messy to use, for one, and many business people felt that carbon copy paper remained perfectly adequate for their needs. Eventually the Battelle Memorial Institute in Colum-

Chester F. Carlson with early photocopier

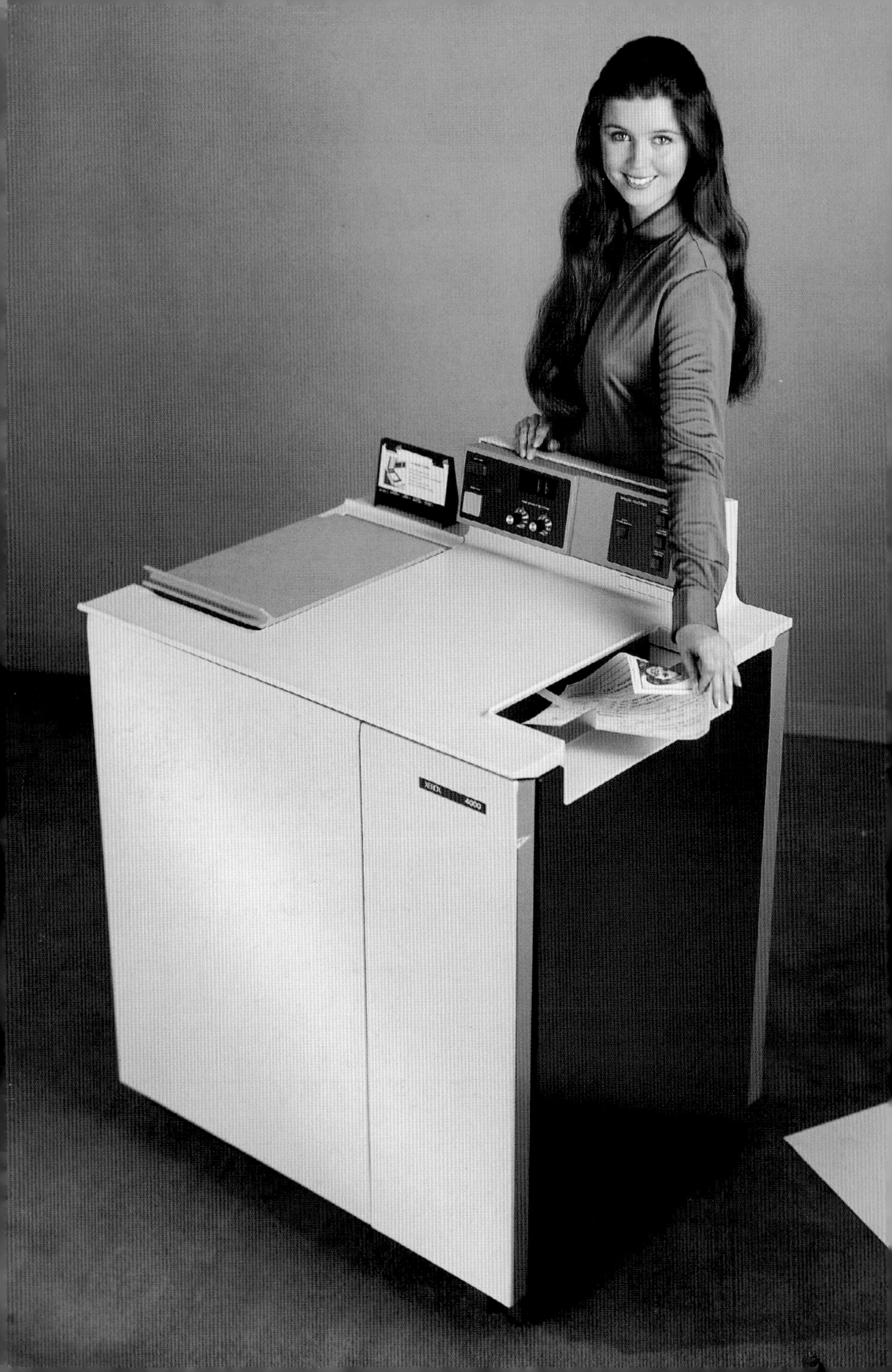
XEROX 4000

Telecopier 400 facsimile machine manufactured by Rank Xerox in England, 1974

-Xerox copier, mid-1970s

bus, Ohio, contracted Carlson to refine his "electrophotography" process in 1944. The Haloid Corporation, a photographic paper manufacturer and retailer founded in 1906, subsequently purchased the rights to manufacture and market a copying machine based on Carlson's invention. Later, the company obtained all the rights pertaining to Carlson's pioneering technology and, with him, agreed that a new and shorter title for the process was needed. The name "Xerography" was eventually coined and the word "Xerox" was trademarked in 1948. The following year saw the introduction of the first xerographic copier, the *Model A*. In 1958 the Haloid Company changed its name to Haloid Xerox Inc. (becoming the Xerox Corporation in 1961). The first automatic plain-paper office copier, the *Xerox 914*, was launched in 1959 and remained in production until 1976. This design was so successful and influential that the company had to battle to prevent the name "Xerox" from becoming a generic term. Since then, Xerox has been a world leader in the photocopying field and has diversified into other product areas, including word-processors, facsimile machines and the highly successful office communications network *Ethernet* (1979). Since the early 1990s, Xerox has been committed to the design of "waste-free" and "sustainable" products that "optimize resource utilization and minimize environmental impact".

LZ 126 airship being repaired, c. 1924

FERDINAND GRAF VON ZEPPELIN

BORN 1838 KONSTANZ, BADEN, GERMANY
DIED 1917 CHARLOTTENBURG, BERLIN, GERMANY

In 1863 Ferdinand Graf von Zeppelin made his first balloon ascent in St. Paul, Minnesota, while acting as a volunteer military observer for the Union Army during the American Civil War. He became fascinated by balloon technology and after retiring from the military in 1890 dedicated his energies to the design of rigid dirigible airships. After ten years of research, Zeppelin's *LZ-1* airship made its first test flight. 128 metres long, it had a lightweight aluminium frame and could achieve speeds of up to 32 kph. The *LZ-1* was the first of many lighter-than-air airships which Zeppelin developed, and which performed better than Germany's fledging aeroplanes. After a Zeppelin airship made a 24-hour flight in 1906, the German government placed an order for an entire fleet. An airship passenger service was launched four years later, by the Deutsche-Luftschiffahrts AG (German Airship Travel Company). Over one hundred Zeppelin airships were used for military purposes during the First World War, after which commercial production resumed under the supervision of Hugo Eckner (1868–1954), who piloted the legendary *Graf*

The *Hindenburg* disaster – the airship crashed and exploded while landing at Lakehurst, New Jersey, on 6 May 1937

upp Wiertz, *2 Days o Europe* poster showing Zeppelin *LZ 129* airship, 1937

Zeppelin on its 21-day around-the-world flight in 1929. When the Nazi Party came to power in 1933, the company was compelled to design an even bigger and better airship that was intended to reflect the superiority of the Third Reich. The result was the *LZ-129*, which made its first flight in 1936. Better known as the *Hindenburg*, this airship famously crashed and exploded while landing at Lakehurst, New Jersey, in 1937. Following this tragedy, airships were deemed too dangerous and all but disappeared from the skies.

APPENDIX

Acknowledgements

We would like to take this opportunity to thank all at TASCHEN for the successful realization of yet another project. We also wish to thank Nick Bell, Sacha Davison and Christopher Brawn of UNA for their wonderful graphic design work and the many individuals, manufacturers, distributors, design offices, auction houses, public institutions and picture libraries who have lent their assistance and provided images. Thanks must also go to Paul Chave for the new photography generated specially for this project.

Picture Credits

We are immensely grateful to those individuals and institutions that have allowed us to reproduce images. We regret in some cases it has not been possible to trace the original copyright holders of photographs from earlier publications. We would also like to thank the numerous designers, manufacturers and institutions that have kindly supplied portrait images. The publisher has endeavoured to respect the rights of third parties and if any rights have been overlooked in individual cases, the mistake will be correspondingly amended where possible. The majority of historic images were sourced from the design archives of Fiell International Ltd., London and TASCHEN GmbH, Cologne.

AEG Aktiengesellschaft 10 (top), 10 (bottom), 11, 12 (top-left), 12 (top-right), 24 (top), 24 (bottom) 25 (bottom), 27 **Emilio Ambasz** 13 (top), 13 (bottom), 14 **Apple Computer** 15 (top), 15 (bottom), 16 (bottom), 17 (top), 114 (top), 114 (bottom), 115 **Archivo Storico Olivetti** 139 (top), 139 (bottom), 140 (all images), 141 **Artemide** 60 (bottom) **Audi** 19 (top), 19 (bottom), 20, 21 (top), 21 (bottom), 22 **Aviation Picture Library (aviationpictures.com)** 39 (bottom) **Bauhaus Archiv** 45 (photo: Gunter Lepkowski) **Biro Bic Ltd.** 31 (top), 31 (bottom), 32 **Black & Decker** 33 (top), 33 (bottom) **BMW Museum** 34 (top), 34 (bottom), 35 (top), 35 (bottom), 36 **Boeing Company** 37 (top), 37 (bottom), 38 (top), 39 (top), 129 (top), 130 (all images) **Braun GmbH** 46 (top), 46 (bottom), 48, 49, 134 (top), 134 (bottom) **BT Archives** 78 (bottom), 155 (bottom) **Buckminster Fuller Archives** 90 (top), 90 (bottom), 91 **Canon Europa NV** 50 (top), 50 (bottom), 51 **Christies Images** 23 (bottom), 167 **Citroën AG** 56 (top), 56 (bottom), 57 (top) **Coope-Hewitt Museum** 70 (bottom) **Corning Inc** 73 (top) **Daimler-Chrysler** 52 (top) 52 (bottom), 53, 54 (top), 54 (bottom), 55, 131 (top), 131 (bottom), 132 (all images), 133 **Deere & Co.** 61 (top), 61 (bottom), 62 (all images) **Design Research Unit** 64 (top), 64 (bottom), 65 **L. M. Ericsson** 78 (top), 79 **Fiat UK** 82 (top), 82 (bottom), 83 **Electrolux** 123, 154 (top) (photo: Hans Thorwid-Nationalmuseum) **Eli Whitney Museum** 179 (top), 179 (bottom), 180 **Fiell International Limited** (all photos by Paul Chave) 47, 127 (bottom) **Fiskars** 84 (top), 84 (bottom) **Fondation pour l'Architecture** 30 (bottom) **Ford Motor Company** 85 (top), 85 (bottom), 86 (all images), 87 (all images), 88 **General Electric Co.** 93 (top), 93 (bottom), 94 (top), 94 (bottom), 95 **General Motors Corp.** 96 (top), 96 (bottom), 97, 98, 99(top), 99 (bottom) **Gillette Company** 100 (top), 100 (bottom-left), 100 (bottom-right), 101 (all images) **GK Design Group** 77 (top), 77 (bottom) **Henrik Cam Phtotography** 80 (bottom), 81 (top) Henry Dreyfuss Associates 69 (bottom), 70 (top), 71 (top), 71 (bottom) 72 **Hoover Company** 107 (all images) **IBM Corp.** 109 (top), 109 (bottom), 110, 136 (bottom), 137 (bottom) **IDEO** 89 (top), 89 (bottom), 111 (top), 111 (bottom) **Kartell** 116 (top), 116 (bottom), 117 **Matsushita** 128 (top) **National Motor Museum (Beaulieu)** 57 (bottom) **Die Neue Sammlung** 30 (top), 59 (bottom – photo: Angela Bröhan), 128 (bottom – photo: Angela Bröhan), 145 (bottom – photo: Angela Bröhan) **Nissan Motor Co.** 135 (top), 135 (bottom) **Oakley Inc** 138 (top), 138 (bottom) **Pentagram Design** 102 (top), 102 (bottom-left), 102 (bottom-right), 103, 121 **Philips NV** 144 **Porsche Cars UK Ltd.** 146 (top), 146 (bottom), 147 **Porsche Design** 148 (top), 148 (bottom), 149 (all images) **Raymond Loewy International** 122 (all images), 124 (all images), 125 (top), 125 (bottom), 126 (all images) **Renault Presse** 152 (top), 152 (bottom), 153 (top), 153 (bottom) **Robert Bosch GmbH** 40 (top), 40 (bottom), 41, 42, 43 (top), 43 (bottom) **Science & Society Picture Library** 28 (bottom), 29, 58, 75 (right), 108, 118 (bottom), 119, 120 (top), 120 (bottom), 129 (bottom), 137 (top), 142 (top), 142 (bottom), 143 (top), 168 (bottom-left), 168 (bottom-right), 181 (top-both portrait images), 182, 185, 186 (top) **Sears, Roebuck & Co.** 158, 159 (top) **Sony Corporation**, 157 (top), 157 (bottom), 159 (bottom) **Springfield Armory Archive** 160 (top), 160 (bottom) **Team Design** 156 (top), 156 (bottom) **Texas Instruments** 164 (top), 164 (bottom), 165 **Torsten Bröhan GmbH** 63 (bottom left) **Tupperware Europe** 4–5, 168 (top), 169 **Vin-Mag** 75 (left), 112 (bottom), 150 (bottom) **Volkswagen AG/Audi** 172 (top), 173 (top), 173 (bottom), 174 (top), 174 (bottom), 175 **The Wedgwood Museum** 176 (top), 176 (bottom), 177 **Werkbund-Archiv** 68 **The Wolfsonian – Florida International University), The Michell Wolfson Jr. Collection** 18 (bottom), 67, 76 (Collection of Finlay B. Matheson), 92 (bottom), 143 (bottom), 151, 161 (bottom), 171, 172 (bottom), 178, 187 **Xerox Corp.** 183 (top), 184

Authors' Biographies

Charlotte J. Fiell studied at the British Institute, Florence, and at Camberwell School of Arts & Crafts, London, where she received a BA (Hons) in the History of Drawing and Printmaking with Material Science. She later trained with Sotheby's Educational Studies, also in London.

Peter M. Fiell trained with Sotheby's Educational Studies in London and later received an MA in Design Studies from Central St Martins College of Art & Design, London.

Together, the Fiells run a design consultancy in London specializing in the sale, acquisition, study and promotion of design artifacts. They have lectured widely, curated a number of exhibitions and written numerous articles and books on design and designers, including TASCHEN's *Charles Rennie Mackintosh*, *William Morris*, *1000 Chairs*, *Design of the 20th Century*, *Industrial Design A–Z*, *Designing the 21st Century*, *Scandinavian Design* and *Graphic Design for the 21st Century*. They have also edited the six-volume *Decorative Art* series published by TASCHEN.

The Fiells can be contacted at: fiell@btinternet.com

Reviews of books by Charlotte and Peter Fiell from TASCHEN

'The latest work from the team of Fiell and Fiell is up to the high standards we have come to expect from these British designophiles."
www.designzine.com, on *Industrial Design*

'Overall, it is an excellent reference and guide for anyone interested in design and creative thinking. Like all TASCHEN books, this book is well written, concise yet articulate; beautifully illustrated and printed."
Reader's comment (www.amazon.com), on *Industrial Design*

'If you take even the slightest interest in the design of your toothbrush, the history behind your washing machine, or he evolution of the telephone, you'll take an even greater interest in this new book. [...] Exploding with color, aesthet-cs, and style, Industrial Design is informative and fun. You'll have a hard time putting it down and it will change the vay you look at everything around you."
Reader's comment (www.amazon.com), on *Industrial Design*

The definite book on the subject."
The Observer Magazine, London, on *Scandinavian Design*

A fabulously detailed reference for design fans that's so chock-full of beautiful pictures it'll tempt anyone with even a assing interest in the subject."
TIME OUT, London, on *Scandinavian Design*

Essential reading for every budding designer."
Australian Interiors, Sydney, on *Designing the 21st Century*

The usual cocktail of excellent quality graphics, photography and typography, inevitably combine to create a visually xplosive book."
Label, London, on *Designing the 21st Century*

A book with exemplary works by 100 outstanding graphic designers has now been published by TASCHEN. There can e no clearer document of what constitutes the 'state of the art'. A fabulous work, it should be on every obligatory eading list."
Designers Digest, Sittensen, on *Graphic Design for the 21st Century*

A sweeping look at some of the most progressive designers."
Blue Print, London, on *Graphic Design for the 21st Century*

A standard reference volume, seen behind the counter of any respectable dealer of vintage 20th century design."
World of Interiors, London, on *1000 Chairs*

Design of the 20th Century
Charlotte & Peter Fiell /
Flexi-cover, Klotz, 768 pp. /
€ 19.99 / $ 29.99 / £ 14.99 /
¥ 3.900

Graphic Design for the 21st Century
Charlotte & Peter Fiell /
Flexi-cover, 640 pp. / € 29.99 /
$ 39.99 / £ 19.99 / ¥ 4.900

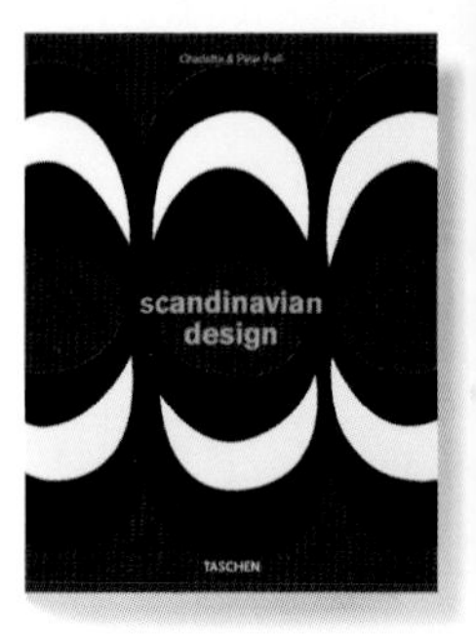

Scandinavian Design
Charlotte & Peter Fiell /
Flexi-cover, 704 pp. / € 29.99 /
$ 39.99 / £ 19.99 / ¥ 4.900

"Fiell Good Factor ... they have turned out a stream of lexicons on 20th century design for the German publisher TASCHEN. Their most recent book is the most adventurous, as it steps outside historical territory to question 100 contemporary designers about their vision of the future."
—*The Saturday Telegraph Magazine*, United Kingdom, on *Designing the 21st Century*

" Buy them all and add some pleasure to your life."

All-American Ads 40s
Ed. Jim Heimann

All-American Ads 50s
Ed. Jim Heimann

Angels
Gilles Néret

Architecture Now!
Ed. Philip Jodidio

Art Now
Eds. Burkhard Riemschneider, Uta Grosenick

Atget's Paris
Ed. Hans Christian Adam

Best of Bizarre
Ed. Eric Kroll

Bizarro Postcards
Ed. Jim Heimann

Karl Blossfeldt
Ed. Hans Christian Adam

California, Here I Come
Ed. Jim Heimann

50s Cars
Ed. Jim Heimann

Chairs
Charlotte & Peter Fiell

Classic Rock Covers
Michael Ochs

Description of Egypt
Ed. Gilles Néret

Design of the 20th Century
Charlotte & Peter Fiell

Design for the 21st Century
Charlotte & Peter Fiell

Dessous
Lingerie as Erotic Weapon
Gilles Néret

Devils
Gilles Néret

Digital Beauties
Ed. Julius Wiedemann

Robert Doisneau
Ed. Jean-Claude Gautrand

Eccentric Style
Ed. Angelika Taschen

Encyclopaedia Anatomica
Museo La Specola, Florence

Erotica 17th–18th Century
From Rembrandt to Fragonard
Gilles Néret

Erotica 19th Century
From Courbet to Gauguin
Gilles Néret

Erotica 20th Century, Vol. I
From Rodin to Picasso
Gilles Néret

Erotica 20th Century, Vol. II
From Dalí to Crumb
Gilles Néret

Future Perfect
Ed. Jim Heimann

The Garden at Eichstätt
Basilius Besler

HR Giger
HR Giger

Indian Style
Ed. Angelika Taschen

Industrial Design
Charlotte and Peter Fiell

Kitchen Kitsch
Ed. Jim Heimann

Krazy Kids' Food
Eds. Steve Roden, Dan Goodsell

London Style
Ed. Angelika Taschen

Male Nudes
David Leddick

Man Ray
Ed. Manfred Heiting

Mexicana
Ed. Jim Heimann

Native Americans
Edward S. Curtis
Ed. Hans Christian Adam

New York Style
Ed. Angelika Taschen

Extra/Ordinary Objects, Vol. I
Ed. Colors Magazine

15th Century Paintings
Rose-Marie and Rainer Hagen

16th Century Paintings
Rose-Marie and Rainer Hagen

Paris-Hollywood
Serge Jacques
Ed. Gilles Néret

Penguin
Frans Lanting

Photo Icons, Vol. I
Hans-Michael Koetzle

Photo Icons, Vol. II
Hans-Michael Koetzle

20th Century Photography
Museum Ludwig Cologne

Pin-Ups
Ed. Burkhard Riemschneider

Giovanni Battista Piranesi
Luigi Ficacci

Provence Style
Ed. Angelika Taschen

Pussy Cats
Gilles Néret

Redouté's Roses
Pierre-Joseph Redouté

Robots and Spaceships
Ed. Teruhisa Kitahara

Seaside Style
Ed. Angelika Taschen

Seba: Natural Curiosities
I. Müsch, R. Willmann, J. Rust

See the World
Ed. Jim Heimann

Eric Stanton
Reunion in Ropes & Other Stories
Ed. Burkhard Riemschneider

Eric Stanton
She Dominates All & Other Stories
Ed. Burkhard Riemschneider

Tattoos
Ed. Henk Schiffmacher

Tuscany Style
Ed. Angelika Taschen

Edward Weston
Ed. Manfred Heiting